Water Supply Systems and Evaluation Methods

Volume I: Water Supply System Concepts

Harry E. Hickey, Ph.D.

Acknowledgment

This project was performed by the Society of Fire Protection Engineers (SFPE) and was supported by the Department of Homeland Security's Science and Technology Directorate and the U.S. Fire Administration (USFA). SFPE is an engineering association for advancing the science and practice of fire protection engineering. Water supply is an important subject to the fire service, fire protection engineers, and city managers. These manuals are intended to provide a reference for concepts and terminology to facilitate communication and understanding between these organizations.

About the Author

Dr. Harry E. Hickey's career in fire protection spans more than 50 years. He taught Fire Protection Engineering at the University of Maryland for 26 years. He also has extensive experience in the fire services both as a firefighter, fire officer and Emergency Coordinator. His combination of municipal fire administration and fire protection engineering experience provides him with unique insight into the challenges of design and operation of municipal water supplies.

He received his Ph.D. in Public Administration from American University in Washington, DC. He has authored many book and articles including *Public Fire Safety, A Systems Approach, Fire Protection Hydraulics,* and two editions of *The Fire Suppression Rating Schedule Handbook.*

Table of Contents

Chapter 1: Fundamental Considerations

Topic 1: Anatomy of a Water System 6

Topic 2: Continuous Availability of Water Supply 11

Topic 3: Municipal Water Supply Considerations 13

Topic 4: Water Sources 16

Topic 5: Classification of Water Systems 19

Chapter 2: Processed Water for Domestic Consumption 20

Topic 1: Unfiltered Surface Water Sources 20

Topic 2: Filtered Surface Water Sources 21

Topic 3: Water Treatment Plants 23

Topic 4: Auxiliary Facilities for Surface Water Treatment 28

Topic 5: Ground Subsurface Water Systems 29

Chapter 3: Water Quality Standards 32

A General Perspective on Water Quality 32

Water Quality Monitoring 39

Chapter 4: Water Distribution System Design Concepts 51

Water Supply Source Classifications 51

Water System Components 52

Water System Classifications 52

Composite Water Supply Systems 56

System Demand, Water Design and Flow Criteria 57

Rates of Water Use 58

Distribution System Appurtenances 58

Hydrant Locations 66

System Evaluation and Design 71

Basic Concepts in Determining Design Flow at System Demand Points 73

Distribution System Storage 75

Elevated and Ground Storage 77

Comparison of System Examples 84

Recommended Practice on Water Storage 84

Chapter 5: Consumer Consumption and Needed Fire Flow 86

Water Demands 86

A Suitable Approach for Studying Consumer Use Water Demand 91

Fire Suppression Water Demand 91

Special Notes on the Determination of Needed Fire Flows 99

Chapter 6: Alternative Water Supplies 101

Alternative Water Supply Concepts 103

Where are Alternative Water Supplies Needed 104

Water Supply Officer 104

Identifying and Certifying Alternative Water Supplies 105

Establishing an Alternative Water Delivery Program 112

Chapter 7: Impacts of Fire Flow on Distribution System Water Quality, Design and Operation 117

Background Statement 117

Background 118

Study Objectives and Scope 118

Chapter 8: Dual Water Systems 138

Basic Concepts 138

Background Information 139

Potential Applications for Dual Distribution Systems 139

Sources of Nonpotable Water 141

Potential Uses of Nonpotable Water 141

Potential Uses for Nonpotable Water 141

Nonpotable Water Reuse Legislation 143

Reclaimed Water Quality and Treatment Requirements 145

Reclaimed Water Monitoring Requirements 146

Treatment Facility Reliability 147

Groundwater Monitoring 147

Treatment for Reclaimed Water...148

Reclamation Plants 149

Storage...150

Features of Reclaimed Water Distribution Systems 151

The Future Role of Dual Water Systems..152

Chapter 9: Water Supply and Effective Fire Protection 153

Overview 153

The Insurance Services Office, Inc...153

The Ability of Fire Departments to Provide Effective Fire Protection.............................154

Fire Insurance Rates 155

An Overview of Water Supply Under the ISO Fire Suppression Rating Schedule 155

Insurance Company Assistance on Evaluating Water Supplies 158

Advantages of Automatic Sprinkler Systems...159

Water Supply Requirements for Sprinklered Properties 161

Design Curves...161

Hazard Classifications..163

Residential Sprinkler Systems...163

Types of Sprinklers for Commercial Buildings 164

Pipe Schedule Systems 164

Hydraulically Calculated Systems 164

Standpipes 165

The Physical Properties of Water 168

Water Mist Fire Protection Systems 171

Class A Foam Fire Protection...174

Chapter 10: Separate Water Systems and Emergency Water Supplies.......................178

Part I...178

A Current Need for Individual Water Systems for Fire Protection 180

Basic Design Concepts for Individual Water Supplies for Fire Protection 181

Part II: Emergency Water Supplies 184

Chapter 1: Fundamental Considerations

Water for human consumption comes from one of two basic sources:

1) Water from a well to supply an individual residence, well water for farmstead properties, and well water for small public sector properties that include schools, public buildings, and small commercial enterprises.

2) Municipal water systems that provide potable water to a wide array of commercial property and domestic use buildings including apartments, condominiums, duplex housing, and single family dwellings.

This chapter uses the Washington, DC, Water System as an example in order to introduce concepts associated with a fundamental understanding of water distribution systems. This system was selected because it typifies many of the water systems in the United States that rely on water sources including rivers, lakes, and low-level water retention dams. Other water supply sources are examined under Topic 3 in this chapter.

Understanding the fundamentals of a municipal water supply delivery system is essential to closely examining the many features of a water system and the many options in designing a water delivery system. Chapter 1 provides a basic overview of:

- The Anatomy of a Water System
- The Need for a Continuously Available Water Supply
- Considerations for Establishing Municipal Water Supply Systems
- Classification of Water Supply Sources
- The Classification of Water Supply Systems

Topic 1: Anatomy of a Water System

The purpose of municipal water delivery systems is to transport potable water from a water treatment facility to residential consumers, for use as drinking water, water for cooking, water for sanitary conditions, and other water use in a domestic environment. Water supply also is essential for business and industry to operate in a municipal environment. Of no less importance is the need to supply water to properly located fire hydrants to provide the public with an effective level of fire protection. Municipal water systems also may need to provide water for special services that include street cleaning, the selling of water to contractors for erecting buildings, parks and recreation, and miscellaneous uses.

A water system has two primary requirements: First, it needs to deliver **adequate** amounts of water to meet consumer consumption requirements plus needed fire flow requirements. Second, the water system needs to be **reliable**; the required amount of water needs to be available 24 hours a day, 365 days a year.

Anatomy may be defined as "separating or dividing a function into parts for detailed examination." (1) A water supply system is analogous to the human circulatory system. The heart pumps blood through the arteries, veins, and capillaries to supply oxygen to all part of the body. A water pump supplies water through primary, secondary, and distributor water mains to supply water to consumers and for fire protection.

This section examines the functional components of the water system by tracking the water from the sources that feed the municipal water system to the water *tap*. The term *tap* is used in a generic sense to mean any reference point on the water distribution piping where a connection or tap is made to supply a lateral pipe to a domestic connection, a commercial connection, or a lateral line to a fire hydrant.

Washington, DC, has a very old water system that has been updated in many ways. It serves as an excellent example of a basic municipal water supply system. Variations to this system are reviewed later in this manual.

Every municipal water system has to have a water supply source that is both adequate and reliable for the city to be served. The primary water source of water for Washington, DC, is the Potomac River. With minor exceptions due to ice jams and flooding, this water supply has been reliable since before the Civil War. Water is fed to the city from intakes at Great Falls and Little Falls both by a gravity aqueduct and by a more modern pumping station. The aqueduct water serves residents of the District of Columbia, Arlington County, Virginia, and Falls Church, Virginia.

Figure 1-1 illustrates a progressive view of the water system. Two holding reservoirs supply water to a treatment plant that processes the water to remove impurities and adds chemicals to bring the water into compliance with the Environmental Protection Agency (EPA) regulations on *clean water* for drinking and commercial cooking. The actual water treatment process is discussed. The purified water, or *finished water*, then is pumped to several different storage tanks and storage basins around the city for release into the distribution system piping network on demand for consumer use or in the case of a working fire. Depending on the different elevations points throughout the city, additional pumping stations are provided to maintain adequate pressure in the water system during varying periods of consumer use or emergency waster supply demand requirements. Water flows from the storage locations through the primary, secondary, and distributor mains to supply service lines to individual water consumers and lateral lines to supply fire hydrants.

This simplified illustration tracks raw water that originated at the Potomac River through the treatment plant, the storage of water, and the movement of water through different sizes of water mains to service outlets throughout the city to supply consumer demands. At the same time, water in the street mains is maintained at adequate pressure to supply fire department pumpers that may need to take water from fire hydrant to suppress structural fires and handle other emergency situations.

Figure 1-2 illustrates eight basic steps that are used to produce a supply of potable water. These specific steps are associated with the treatment plant for the city of Washington, DC, and fit into the treatment facility portion of **Figure 1-1**. While this is a generally acceptable method of purifying water, it should be recognized that every water supply has its own specific requirements for establishing potability. Chapters 2 and 3 continue the discussion of processing water for consumer use and establishing water quality standards by the EPA.

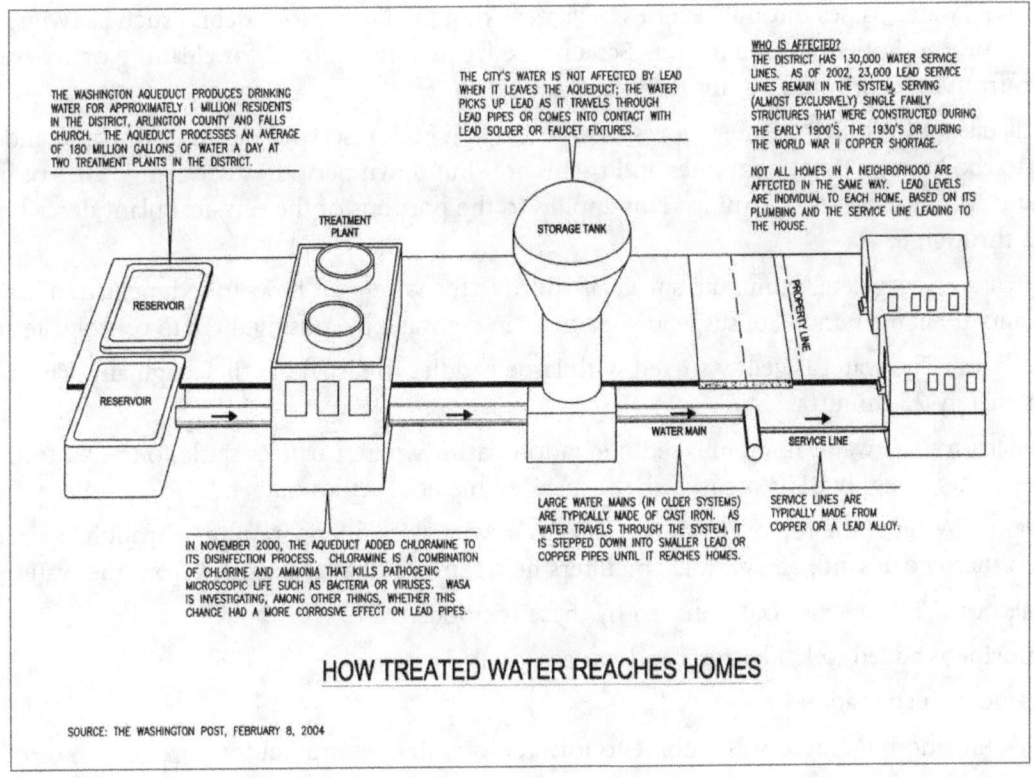

Figure 1-1: Features of a Small Community Water Distribution System

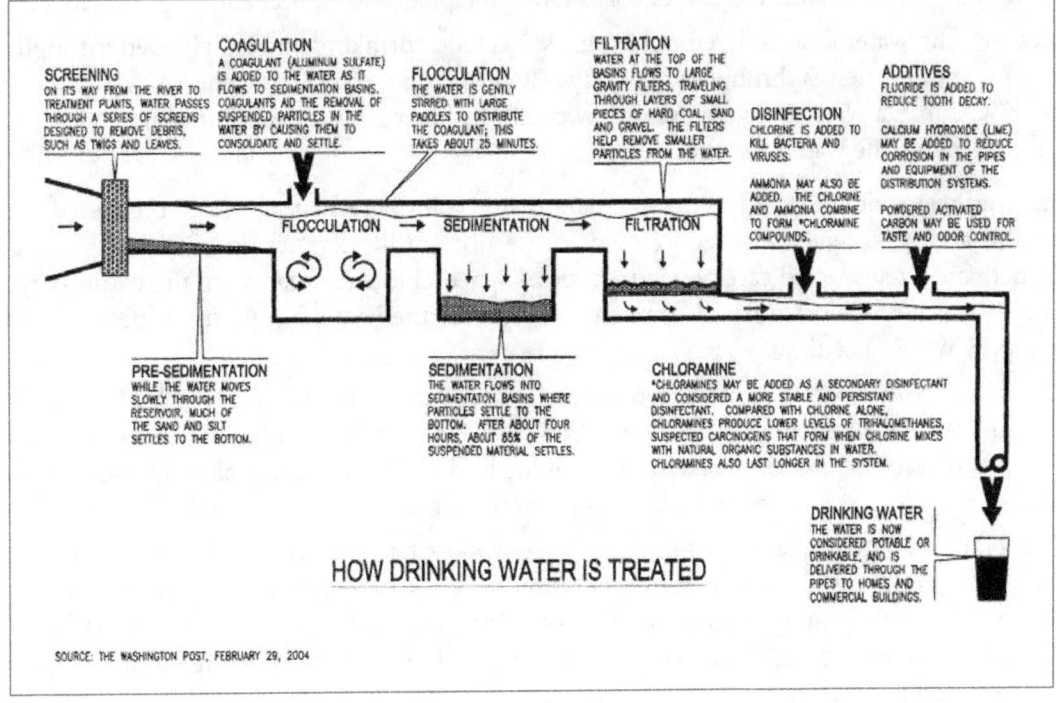

Figure 1-2: Features of a Minimum Size Community Water Distribution System

Step 1: *Screening* - Water passes through a series of screens designed to remove debris such as twigs, leaves, paper, stones, and other foreign matter. Screens are frequently removed for cleaning or are back-washed from high-pressure pumps to prevent clogging.

Step 2: *Presedimentation* - While the water moves slowly through each reservoir, much of the sand and silt settles to the bottom. Treatment lines and basins are shut down periodically during times of minimum domestic consumption for cleaning. This applies to the portions of the physical plant described in Steps 2 through 8.

Step 3: *Coagulation* - A coagulant, aluminum sulfate, is added to the water as it flows to sedimentation basins. Coagulants aid in the removal of suspended particles in the water by causing them to consolidate and settle.

Step 4: *Flocculation* - The water is gently stirred with large paddles to distribute the coagulant. This takes approximately 25 minutes.

Step 5: *Sedimentation* - The water flows into sedimentation basins where particles settle to the bottom. After about 4 hours, roughly 85 percent of the suspended material settles out.

Step 6: *Filtration* - Water at the top of the basins flow to large gravity filters, traveling through layers of small pieces of hard coal, sand, and gravel. The filters help remove smaller particles from the water.

Step 7: *Disinfection* - This may be accomplished by these methods:

  ~ Chlorine is added to kill bacteria and viruses.

  ~ Ammonia also is added.

  ~ The chlorine and ammonia combine to form chloramines compounds.

Step 8: *Additives* - Depending on the quality of the water at this point, the following additives may be injected into the water stream to accomplish the stated benefits:

  ~ Fluoride is added to reduce tooth decay.

  ~ Calcium hydroxide is added to reduce corrosion in the pipes and equipment of the distribution system.

After processing, the water is considered potable or suitable for drinking and is delivered through the pipes to homes and businesses throughout the city. However, due to special conditions associated with the District of Columbia Water Delivery System, two other additives have been introduced into the water supply before it enters the water distribution system:

1) Chloramines were added recently as a secondary disinfectant because they were considered to be more stable and persistent compared to the chlorine alone. Chloramines produce lower levels of trihalomethanes, a suspected carcinogen that forms when chlorine mixes with the natural organic substances in water. Chloramines also are favored because the have a longer use expectancy time in the water supply without settling out.

2) The District of Columbia has experienced a severe problem of excessive traces of lead in the finished water from old lead-lined pipes in portions of the city. To counteract this problem, orthophosphates have been added to the drinking water to reduce the lead level. Orthophosphates form a protective coating on pipes to prevent lead from leaching into the water.

There are several different adaptations of the components of water systems in order to provide adequate water pressure throughout the entire water distribution system and to assure an adequate quantity of water to meet domestic consumption needs, fire protection needs, and special-use needs. Different arrangements for supplying municipal water, and the arrangements of water storage in municipal water systems are explored in Chapter 4.

TOPIC 2: **Continuous Availability of Water Supply**

A municipal water supply system cannot service its customers unless there is a continuous supply of water to meet domestic consumption needs in the broadest sense and water needs for structural fire protection. Water sources need to be selected carefully to make sure that this fundamental requirement is met. Two main factors that affect water supply selection are

1) Quality of water: Water must be treated or purified to meet Regulatory Requirements established by the EPA (United States Government). The requirements are divided into 2 categories:

   a. residential communities with populations not exceeding 3,000; and

   b. combined residential and commercial communities that serve a population demand over 3,000.

   Water quality standards are covered in Chapter 3.

2) Quantity of water: The quantity of water must be adequate to meet consumer consumption and fire-flow demands at any time of the day, day of week, and week of the year.

Maintaining a continuous or uninterrupted supply of water for municipal demands is a major challenge to many municipalities because of the following conditions:

- droughts;
- growing demands that cannot be met by the treatment plant;
- lack of adequate storage capacity;
- other communities drawing water from the same supply sources such as a lake or a river;
- a major commercial fire or wild land/urban interface fire that exhausts the water supply; and
- undetected underground leakage on the pipe distribution system.

A municipality must recognize that the quantity of available water needs to be such that maximum daily consumption demands are satisfied at all times, even during periods of drought or after years of community growth. The water delivery system needs to expand as the municipality expands.

Design Considerations

In the design and construction of community water systems, economics are extremely important. This dictates that the source of supply should be selected so that little maintenance for the operational factors will be required to furnish an adequate supply of water to the community. Even though surface water supplies such as lakes and rivers, with proper treatments, are by far the more suitable from the standpoint of adequacy, the use of surface water supply systems is confined to municipalities that have a water demand in excess of 75,000 gallons per day domestic consumption to be economically feasible. This translates to a community of about 300 residents with 12 commercial establishments, but with no manufacturing, and the ability to supply a fire flow of 500 gallons per minute (gpm) for 2 hours. This also means that ground water sources are suitable only for quite small communities typically found in more rural areas of the country.

There is an important exception to the above conclusions that ground water sources typically are limited to the very small communities. This assumes that a well site or a few well sites use well pumps to supply water directly into the distribution system after water treatment. Ground water sources can be used to pump water directly into holding basins, small reservoirs, ground-level storage tanks, or even elevated storage tanks where the amount of treated water in storage governs both the population served and the available water supply for fire protection. A southern city, which shall remain nameless, with a population of approximately 75,000 obtains all its water from well sites and then pumps this water into storage which then supplies the distribution system upon demand. However, the price per thousand gallons of finished water is substantially higher than where surface water is used as a community water supply.

Ground Water

In the original planning of ground water supplies, little can be done about determining the chemical quality of the water because the water will be obtained from several well-defined and different water-bearing geological layers or strata. The chemical or mineral quality of the water contributed from each of these water-bearing formations or *aquifers* will be dependent on the dissolution of material within the formation. Therefore, water withdrawn from any ground water source will be a composite of these individual aquifers. The water quality can be determined by actual sampling and analysis of the completed wells. Local and State health departments, along with State departments of environmental conservation (DECs), will have general and helpful information on water quality within their jurisdiction from years of test data. Most State geologists and hydrologists also will have information on both quantity and quality of water within certain regions of their State. All such information is valuable, and should be obtained when considering ground water sources as either primary or secondary supply sources.

Ground water generally will be considered the most readily available source of water because it can be tapped from below the water table beneath the earth. In contrast, surface waters may not be readily available. In many cases, impounding reservoirs provide adequate water for large communities. The water then must be piped some distance to service the community. The quality of the water also may need to be considered when investigating surface waters because of pollution, which may render available water unusable for potable water supplies, even if complete treatment is provided.

Ground water is an integral part of the hydrological cycle of rain and evaporation of water between the earth's bodies of water and the vast moisture content of the atmosphere. This portion of the water cycle

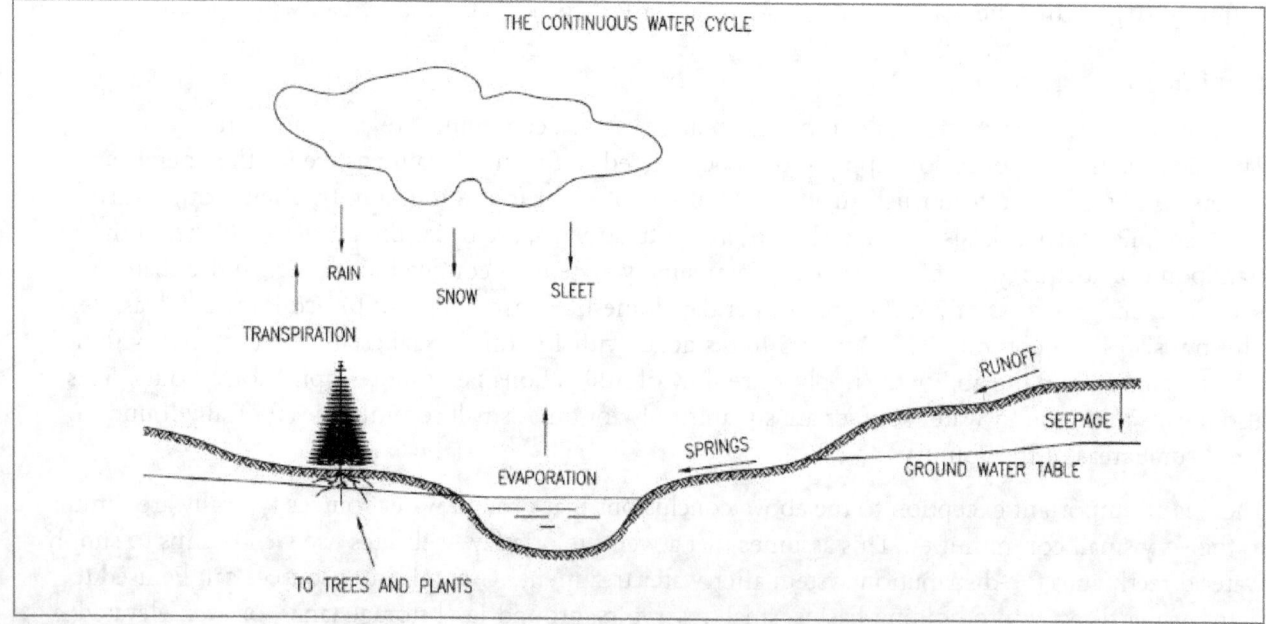

Figure 1-3: Typical Small City Distribution System

is illustrated in **Figure 1-3**. The water content of the atmosphere precipitates from small nuclei and falls to the earth in the form of rain, sleet, and snow. Approximately 70 percent of this precipitation finds its way into the streams, rivers, and finally to the oceans. The remaining 30 percent seeps into the earth and joins the ground water table. Some of this water returns as surface water from springs, flowing wells, etc. Some is used by plants and trees. The remaining portion joins the ground water and appears in layers known as aquifers. The seepage of this water through the earth depends on properties of the soil and rock formations under the earth's surface. Seepage is much greater through porous formations than through the impervious formations. Water seeps at much greater rates through fissures or cracks within the underlying formations. Ground water in the aquifers flows from higher elevations to lower elevations.

In the continuous cycle of water transportation from the earth to the atmosphere, water leaves the earth by evaporation from bodies of water and transpiration from plant life, and returns to the atmosphere where it is retained until again released in the form of rain, sleet, or snow. Replenishment of the ground water table may be accomplished to some extent by the location of ponds and lakes on watersheds. This phenomenon, relating to ponds and lakes, is reversible, as it supplements the ground water table during periods of drought and supplements sustaining flows during the wet season.

TOPIC 3: Municipal Water Supply Considerations

There are two fundamental considerations for both designing and evaluating municipal water supply systems. The first and most important is the quality of the water for human consumption—*drinking water*. The second is the quantity of water required.

In recent years the standards for water quality have been transferred from State health control agencies to the Federal government through two organizations:

1) The United Sates Public Health Service.

2) The EPA.

Water quality is the subject of Chapter 3. Some foundation concepts that relate to the other topic of Chapter 1 will be covered.

The quality of water provided by a municipal water system is based on three distinct characteristics, each of which may independently govern the desirable portability of the water. These characteristics are

1) **Physical quality of water**: The physical quality of water is the *appearance* of the water to the consumer. Physical quality includes the clearness of the water, taste, odor, and temperature. For water to be of attractive physical quality, it must be clear in appearance, or have low turbidity (less than 5.0 units of turbidity). The color of the water must be low in concentration so as not to distract the consumer's attention. Color should be less than 15.0 units of color.

The water should be free of substances that may produce taste and odors upon the addition of chlorine, or upon use of water for cooking purposes. It also should be free of trouble-producing organisms such as aromatic oils of algae or higher bacteria.

The temperature of the water will affect the attractiveness to the extent that use by consumers will decrease if the water is of extremely high temperature. Ground water temperatures vary slightly from around 40 to 55 °F (4 to 13 °C). Such temperature changes are dependent upon well depth and aboveground storage facilities. Surface water temperatures vary with seasonal change from around 40 to 80 °F (4 to 27 °C) with even higher temperatures in the deep South and Southwest.

2) **Bacterial quality of water**: The most important quality of water is that of bacteria content. In the early 20th century, disease outbreaks from water and food-borne bacteria were common throughout the world. Progress in bacteriology and water treatment engineering has all but eliminated outbreaks of water-borne communicable diseases in the United States. Chapters 2 and 3 address this issue in detail.

3) **Water chemistry**: Water is an excellent solvent, so it is not surprising that it picks up other chemicals. During this cycle of water movement, water picks up many solid and gaseous components. As the raindrops fall to the earth, they absorb gases. Most gases within the atmosphere are carbon, sulphur, and nitrogen compounds. The raindrops may also pick up particulate materials in the atmosphere. Many of the particulates are soluble in water and will dissolve within the raindrop.

Other constituents are added to the water cycle from surface or ground water flow. Many and varied constituents are added to the water from dissolution of rocks and minerals which come in contact with the water and its movement. Of particular importance to the water supplier are the following constituents:

~ acidity and alkalinity;

~ calcium;

~ carbon compounds;

~ chlorides;

~ fluorides;

~ iron;

~ magnesium;

~ manganese;

~ nitrogen compounds;

~ silica; and

~ sulphur compounds.

All of the above and others dissolved in water, determine the chemical quality of the water. Each of these constituents has threshold limits that are governed by the EPA, as discussed later.

Municipal Water System Demands

The demand for water supplied by a municipal water system has two driving components: 1) consumer consumption: the amount of water in gpm or gallons per day that is used by all of the taps on the water mains to supply single-family homes, multiple-family residences of all types, health care facilities, schools at all levels of education, commercial enterprises, industrial complexes, and adjunct uses (street cleaning; water fountains; watering public grass areas; shrubs, trees, and flowers; parks and recreation including swimming pools; and the sale of water to contractors for building roads, structures, etc.) and 2) an adequate and reliable water supply for fire protection. Each of these topics is considered in length in future chapters. However, some important terms associated with water demand need to be identified and operationally defined at this time because these terms will be used throughout both the *Concepts Manual* and the *Evaluation Manual* on Municipal Water Supply Systems.

Consumer consumption: Consumer consumption is assessed by determining the amount of water that actually is used by consumers, based on three levels of usage as follows:

1) **Average daily consumption (ADC)**: Each municipal water system services a defined population as determined by census figures. For example, assume that the municipality in question has a population of 22,570. It was determined from water meter readings that the average daily consumption was 137 gallons per person per day or the community consumed 22,570 × 137 gallons = 3,092,090 gallons. This means that on an average day, the water supply works need to have available nearly 3.1 million gallons of water over a 24-hour period, or an average delivery rate of 2,147 gpm of finished water delivered into the water supply distribution system.

   An ADC rate of 137 gallons per day per person is just below the 2003 American Water Works Association (AWWA) average of 141 gallons per day per person. The number reflects the total amount of water used in a day by the population and does not consider usage by different classes of occupancy including commerce and industry. AWWA reports this figure varies considerably by State and region. This information will be broken down in detail in Chapter 5.

2) **Maximum Daily Consumption (MDC)**: This value represents the single day within a year-long period on which the consumption rate was the highest. AWWA reports that the MDC rate for any given community is approximately 150 percent of the ADC rate. For the average community in the United States, this would be 221 gallons per day per person. The MDC rate often is reached in the summer months or during times of peak water demand for industrial use. For example, a company in a community of about 14,000 in western New York produces ketchup for a major retailer during the tomato season. In this case, the MDC rate went to 420 gallons per day per person, a rate of 4,083 gpm at the peak delivery time into the distribution system. When the plant was working at full capacity, a rate of only 130 gpm was available in the center of the municipality for fire protection.

3) **Instantaneous flow demand**: There are generally two peak periods in the day when consumption is greatest: between 7 a.m. to 9 a.m. and between 5 p.m. to 7 p.m.

   The water supply superintendent, or a person of equal responsibility, has to predict these rates in order to control the amount of water delivered to the water distribution system and water pressure such that any given tap can supply water at the desired pressure.

Fire flow demand: At any time, the municipal water supply system should be able to deliver needed fire flows to representative fire risks throughout the municipality from properly located fire hydrants.

An adequate amount of water is essential to confining, controlling, and extinguishing hostile fires in structures. The actual amount of water needed differs throughout a municipality, based on different building and occupant conditions. Therefore, water damage for structural fire protection must be determined at a number of different locations throughout a given municipality or fire protection district. These locations are selected by the Insurance Services Office, Inc. (ISO), to represent typical fire risks, including residential, commercial, mercantile, institutional, and industrial properties for insurance rating purposes.

According to the ISO, the minimum creditable water supply is 250 gpm for 2 hours or a total water supply of 30,000 gallons. Most residential occupancies have a minimum water requirement of 500 gpm, and commercial properties can range up to 12,000 gpm for 4 hours. Chapter 6 covers this subject in detail.

TOPIC 4: Water Sources

Before moving forward with specific details of municipal water systems, we will identify the primary sources of water available for use by municipal water systems and the adequacy and reliability of each. Many municipal water systems use more then one type of water source.

Impounding Reservoirs

Reservoirs created by placing dams across rivers, streams, or at the neck of a valley to capture runoff have been a source of North American municipal water supplies since the mid-1600s. The Grand Coulee Dam on the Columbia River provides water not only to municipal water systems but also to irrigation systems. Reservoir systems are especially common in the northern and northeastern United States. A dam just 200 feet high at the overflow point can create a pressure of nearly 87 pounds per square inch (psi) pressure at the base of the dam. In many cases, this permits the water to flow through the lower portion of the dam to the treatment plant without any pumping requirements.

[Sidebar: Basic concept of hydraulics applied to water supply systems will be covered in *The Manual on Evaluation Municipal Water Supply Systems*.]

In Topic 1, an analogy was made between a water system and the circulatory system of a human being, with the water pump serving as the heart of a municipal water system. In the case of properly designed reservoirs, the head of water creates a pressure to move the water not only through the treatment plant but also through the water distribution system. This concept is illustrated in **Figure 1-1**. Many communities in the United States were supplied with water without pumping stations until after World War II, when increasing demand required the installation of pumps to maintain needed water pressure on the system or to pump water up to grading tanks on the water system that then would supply water by gravity feed to portions of the municipality.

Impounding reservoir systems without a mechanical interface are considered the most reliable and economical source of municipal water supply. More recently constructed water supply systems use this concept through the use of standpipe tanks or gravity tanks on water systems when there is a minimum demand on the system. When there is a higher demand, the water flows by gravity feed from these holding tanks to meet the system demand. Standpipe tanks and gravity tanks will be covered in a subsequent chapter.

Fresh-Water Lakes

The best examples of fresh-water lakes for supplying municipal water come from communities surrounding the Great Lakes, including Chicago, Detroit, Cleveland, Erie, PA, Buffalo, NY, and Rochester, NY. These lakes are supplied by countless streams and rivers that, in turn, are supplied from water runoff from hills, especially during the winter and spring seasons. This is generally considered a functionally, inexhaustible supply of water, provided the intake lines are below the ice formation level, in order to prevent jamming at the inlet ports of the water supply system.

Inland fresh water lakes, such as the Finger Lakes in New York, also supply water to many municipalities. Manmade lakes may be an important solution for coping with increasing water demand.

Without question, rivers across the country provide the most primary sources of water for municipal water systems. The Colorado, the Missouri, the Mississippi, the Ohio, the Tenessee, the Hudson, the Susquehanna, the Potomac; the list could go on and on. However, water shortages are reported on some of these rivers because of the reduction in snowfall and rainfall during periods of the year. Several articles call attention to the radical reduction in the waterway of the Missouri River of late, and the fact that some municipalities have actually run out of water at their intake locations.

Side water damming is one approach to retaining water during dry spells. Other approaches to this problem include cross-connecting rivers and lakes with aqueducts that could be opened on demand to move water from one region to another. There even has been consideration of a transcontinental aqueduct that could be fed from the North to supply the South, the East, and the West. When desalination becomes economically feasible, this aqueduct could be supplied from both oceans.

Wells

Driven or bored wells are excavated to the level of the water table. Artesian or flow wells are a special classification of wells in which the water is trapped below a rock formation, which causes the water to be under pressure of varying intensity. When a tap is made through the bedrock, the water flows to the surface under pressure. This pressure may be sufficient to move the water to the treatment plant. All other types of well require low-lift or high-lift pumps depending on the depth of the water or the distance of the water from the surface.

Wells are the primary source of water for municipalities with populations up to around 5,000. Larger municipalities may use a series of well fields, and usually pump up to gravity tanks to provide the required flow and pressure to the distribution system. **Figure 1-1** illustrated a basic well site, treatment plant, and elevated storage of the finished water.

In many parts of the country, especially on the east coast, the underground supply of water is not being replenished as quickly as it is being consumed. At the end of 2005, this was a special problem in the States of New York, Pennsylvania, Ohio, Virginia, and Florida. In some cases, water is being transported by tanker trucks to meet the minimum needs for domestic consumption. Alternative water sources such as streams, rivers, ponds, and lakes are the only source of water supplies for mobile tankers to provide fire protection. This subject will be covered in a later chapter.

Oceans and Bays

Until recently, the salinity of the water has made this source unacceptable for human consumption and therefore, not a source for municipal water supplies. New technology is now making desalination cost-effective.

Water systems using ocean and bay water for fire protection have been used on both the east coast and west coast for over a century. Pumping stations provide both low-pressure and high-pressure water supplies to specially marked fire hydrants, primarily in the commercial and industrial districts of large cities like New York, Philadelphia, and San Francisco. Some coastal communities in Florida make use of limited coverage fire protection systems that take water from the Atlantic Ocean or the Gulf of Mexico.

TOPIC 5: Classification of Water Systems

Civil engineers have developed three classifications for water supplies:

1) **Fresh water**: This implies that the water source comes from either the surface of the earth or from the natural runoff of water through the water cycle discussed in Topic 2. By definition, these waters are not "brackish," are not to be considered "polluted," and have no trace of salt. This category includes only those water supplies that can be treated by standard methods covered in Chapter 3 to meet EPA criteria for human consumption. Availability of fresh water supplies is becoming an ever-increasing problem throughout the United States. What this means to the consumer is that, like almost every other commodity, the unit cost of finished water is rising, more in some areas of the country then in others. Municipalities need to enforce the Federal government's Clean Water Act.

2) **Salt water**: As stated above, salt water traditionally has not been used as a source of water for domestic consumption. However, recent developments in the reverse osmosis and filtering technology allow the elimination of almost all salt content from ocean waters. The military has used this basic technology aboard ships and through special army units since the Vietnam era, but it has not been cost-effective for municipal water systems.

The latest technology is in evidence in the Tampa Bay, Florida, municipal water system which now satisfies 60 percent of the consumer use needs from a salt water source. This technology has great promise for coastal areas of the country and possibly for transport to interior regions of the country. Cape May, NJ, opened two desalination plants in July 1998. This is certainly the wave of the future for many, many municipal water supply systems.

3) **Reclaimed water**: Sewer water also is required by health regulation to be filtered and treated before it can be discharged into lakes, rivers, and landfills. Once this process has been completed it can be "reclaimed" by going through another treatment plant process for re-entry into a municipal water system. Although technology is available, the cost is unacceptable for most cities. However, scientific work still is progressing in this area.

CHAPTER 2: Processed Water for Domestic Consumption

SURFACE WATER SOURCES

Surface water sources are derived directly from stream and river flow, or are stored prior to use, usually from behind high- or low-level dams that form water retention lakes anywhere from a few acres to many square miles in size. Factors such as chemical and bacterial quality greatly influence the economics of water treatment and the physical quality of the water.

Surface water supplies are divided into two distinct classifications, filtered and unfiltered. These classifications are based upon the type of treatment necessary to produce potable water, and upon the quality of such water prior to any required treatment process.

TOPIC 1: Unfiltered Surface Water Sources

In many instances, a water supply is delivered from a watershed area that is entirely owned or completely controlled by the water company or water authority.

The enforcement of specific rules and regulations is necessary to ensure water quality with an unfiltered water supply. All types of habitation, cultivation, recreation, or other nonwatershed-related use need to be prohibited. In addition, forest fires could create widespread erosion problems, making the water unsuitable unless complete treatment in the form of coagulation and filtration is used. Therefore, a well-planned forestry program including timbering operations, thinning, reforestation, and maintenance of fire lanes throughout the entire watershed area should be inaugurated under the supervision of a trained forester.

All such operational plans should include location and construction of logging roads, repair of equipment, refueling of equipment, and sanitation procedures for workers within the watershed area. Care should be taken so that the discharge of greases, oils, and other fuels is avoided to the fullest extent possible. Sanitation rules involving the disposal of excreta and garbage should be strictly enforced. Central stations, for use by workers, for the disposal of excreta and garbage should be established, preferably below the water works intake. If such stations are located above the intake works, extreme care should be taken in their location and their use should be rigidly enforced.

The entire watershed area should be posted and properly patrolled. The frequency of inspections should depend upon the potential hazard for pollution. If the area is more isolated, the frequency may be monthly; otherwise, constant patrolling should be established. Required water treatment facilities must be adequately supervised on a daily basis or more frequently as deemed necessary!

Treatment of water derived from such a controlled watershed usually consists of coarse screening and continuous chlorination. Preferred treatment would consist of fine screening, pressure sand filtration, pH adjustment, corrosion control, and continuous chlorination.

Watershed sizes and yields will be discussed in the following section, applicable to both unfiltered and filtered supplies.

TOPIC 2: Filtered Surface Water Sources

Surface water sources requiring complete treatment include those that are not entirely owned, supervised, or controlled by the water company or authority. The water will contain normal bacteria content commonly associated to the community life; this excludes gross pollution from sewage, industrial waste, or additives to "spoil" the water. Given these exclusions, proper treatment will render the water potable.

Present EPA bacterial standards indicate that the bacterial load should be less than 5,000 coliforms per 100 milliliters. Present-day loadings indicate that complete treatment of the water, providing coagulation and super-chlorination, will capably treat bacterial loading much greater then this. Water from human-inhabited environments generally will contain high turbidity and possibly some taste- and odor-producing substances.

The final decision in the selection of a source of water for a surface supply should balance consideration of both water quality and water quantity. Water quality includes such factors as bacteria, suitability for treatment, taste- and odor-producing substances, and the effect of further land use along the tributary streams from which the water is to be taken. The source should be adequate for many years into the future, taking into account expected community growth.

In the balance of quality and quantity of water sources, it is desirable to use the largest watershed or resource possible. However, this is not possible in many cases because of the existing pollution of such a source. Therefore, to reach this balance, smaller watershed areas are developed by means of raw water impoundments, often with low-level dams, thereby meeting the requirements of an average water quality with a better-than-average water yield.

The pollution problem of large water sources is becoming more apparent. These larger bodies of water are more suitable for receiving sewage and industrial waste because of dilution and oxygen assimilating capacity. From the polytonal standpoint, the water quality of smaller lakes, rivers, and streams probably will be the most suitable for municipal water supplies in the future.

There are several methods of evaluating the yield of a watershed, storage requirements:

- The MASS Diagram allows the use of past records of rainfall and runoff, gauging station measurements of stream runoff. On the MASS Diagram, the accumulated stream measurement is plotted versus time. On the same diagram, the maximum daily draft is plotted in the same manner (as accumulated values). The required amount of storage is equivalent to the largest depression or dip within the curve, plotted from runoff and as measured down from a parallel of draft.

- Applying storage requirement information from other water management system companies within the same catchment basin. Such information should include a draft, yield of watershed per square mile, and adequacy of present storage, with figures indicating draw-down of these facilities during extended drought periods.

♦ With knowledge of average yield of a watershed, the maximum daily draft of the community supply provides a reservoir with storage of 180 days' capacity with additional capacity for silting and evaporation.

Using a simplified method with a combination of the above methods, along with accepted practice and experience, it is possible to derive certain simplified equations that will provide reliable information concerning watershed yield, maximum daily draft, and the required storage for known watershed areas and predesignated drafts. These drafts may be stated as follows:

From un-impounded streams or rivers where the water is taken directed from the flowing water, by means of a small diversion dam, the draft will depend directly upon the minimum stream flow.

Experience over the years has indicated that the minimum dependable yields for watershed areas will be approximately within the values in **Table 2-1**.

Table 2-1

Average Annual Rainfall: Inches per Year	Minimum Dependable Yield: MGD per Square Mile*
30	0.0055
32	0.0062
34	0.0075
36	0.0088
40	0.012
42	0.014
44	0.016
46	0.020
48	0.023
50	0.026
52	0.030
54	0.035
56	0.040
58	0.048
60	0.056

*MGD = Million Gallons per Day

Dependable draft rate (DD) can be calculated based on watershed area and yield. The water system should be designed so the water source will be dependable for a long period of time. If the area of the watershed is 27 square miles (and the annual rainfall is 46 inches per year), what would be the dependable draft available from the retention area of the watershed?

Dependable Draft (DD) = Area of Watershed (Wi) × Minimum Yield of Watershed (Y)
(MGD)   (Square Miles)   (MGD) per Square Miles Table 2.1

A simplified formula can be used for the determination of the amount of storage required to supply certain daily requirements from a known watershed area. This formula relates back to the previous operation for minimum dependable yield of watershed areas based on average rainfall data. An impoundment should not be expected to yield an average daily amount greater than 25 times the minimum dependable yield of the watershed area. The following simplified formula is applied to relate draft and watershed to area reservoir volume.

Volume of Storage Required in Million Gallons* $=$ Draft in MGD $-$ Minimum Dependable Yield (MGD) per Square Mile $\times$ Area of Watershed In Square Miles

$$V_r = 4.68 \times (24 - 0.02 \times 72)^2$$

$$= 2,370 \text{ Million Gallons}$$

It is possible to calculate the maximum actual draft for an existing reservoir using the same equation, if the volume of storage, the minimum dependable yield of the watershed, and the watershed area are known.

$$DACT = DD + (0.214V_r)^{1/2}$$

or

$$DACT = (W_s \times Y) + (0.214V_r)^{1/2}$$

V_r = Reservoir Volume

DACT = Actual Draft

DD = Dependable Draft $= W_s \times Y$

For example, given the area of the watershed as 75 square miles and an average rainfall of 46 inches per year, the amount of storage can be calculated using the above equation with the values from **Table 2-1**.

TOPIC 3: **Water Treatment Plants**

The treatment facilities for surface water sources are more complex than ground water treatment facilities in both design and operational standpoints, because of the lower quality of the raw water to be treated. Such treatment facilities generally include the following components:

1) **Intake works:** The intake works are structures that are used to divert and lift the water from a stream or impoundment to the treatment works. Intake structures may be concrete in the form of towers with multiple ports located at different elevations, or a crib weighted down within a stream channel. When intakes are located in stream channels, it is important to protect them from floodwaters or damage from floating debris. If possible, the intake structure should be located in the quiescent portion of a stream rather than the highly turbulent portion. The location and design of the structure will minimize the entrance of sand, silt, fish, and debris. Rough or course screening is necessary, and openings should be provided with bars spaced from 1 to 3 inches apart, with the area of the openings restricting the entrance velocity to less than 20 feet per minute. If a diversion dam is used within the stream channel, provisions should be made for the removal of silt and sediment from behind the dam.

   All raw water pumping facilities should be provided in duplicate. Standby or auxiliary gasoline- or diesel-driven pumps also should be provided for large installations. For proper maintenance of equipment, all-weather roads should serve all installations easily.

   The intake structure for raw water impoundment should be located within the deeper portions of the reservoir. Inlets for the intake structure should be located at various levels, depending on the mean and low water level and depth of the water at this location. Surface water is subject to algae growths, whereas the lower waters carry organic matter, silt, and high concentrations of iron, manganese, and sulphide. Course screening or gratings should be provided and designed as mentioned to satisfy stream intakes.

All piping or tunnels for intake structures should be designed for water velocities of 2 to 4 feet per second in order to be self-cleaning. Fine screens are desirable for the protection of water pumps from grit and other debris.

2) **Raw water impoundments:** The capacity of the raw water reservoir should be calculated in accordance with the material presented previously. Certain basic requirements need to be followed concerning the preparation of the reservoir:

   a) An adequate amount of land should be purchased in addition to that acquired for the water area so that proper controls may be put into place regarding the prevention of pollution from use of the land along the water's edge. The additional strip of land should be measured from the high water margin of the reservoir. It will also be used for patrol and reservoir maintenance programs.

   b) An area of 2 feet above the high water margin and 5 feet below the mean level of the reservoir should be cleared and grubbed of all growth. This will prevent the creation of an environment for breeding insects.

   c) The entire area below the 5 foot level from mean water elevation should be cleared and grubbed of all growth. It may be desirable to cut off stumps of 6" and larger at ground level. All such material should be burned or removed from the reservoir area to prevent the presence of organic decay, which leads to taste and odor problems. Care should be taken in the planning of the clearing and grubbing program for large reservoirs. The long period of time involved in construction of the dam, spillway, and intake structure will provide ample time for the vegetation to regrow, necessitating clearing and grubbing the reservoir a second time.

   d) Of extreme importance is the preparation of the areas where the tributary streams enter the reservoir, as these areas usually will be somewhat swampy. Therefore, islands should be constructed to give deeper water areas **or** the material dredged entirely from the reservoir. In many cases, small impoundments are constructed within these areas so that it is possible to control the breeding of mosquitoes by fluctuation of the water level within the small impoundment. The small impoundment must be constructed with the toe of the dam within the larger reservoir in order to prevent the creation of a shallow water area. It will also serve a secondary purpose in that not only will it provide some additional storage, but it also will act as a settling basin for incoming silt from the watershed area.

   e) **Raw water pipelines:** All pipes or tunnels used for intake conduits should be designed for water velocities of from 2 to 4 feet per second in order to be self- cleaning. An adequate number of blowoff valves and air relief valves should be provided along all pipelines for maintenance purposes.

3) **Water treatment plants:** The water treatment plant provides chemical treatment in the form of coagulation of suspended solids, screening or filtering, corrosion control, and chlorination. These processes result in high quality, potable water. The principle of operation is that the water enters the plant; chemicals are added; mixing, sedimentation, and filtration follow; then final adjustment of the water is provided for corrosion control and disinfection. Details of design of these appurtenances are as follows:

Mixing process: One of the most important processes in water treatment is the mixing process whereby the added chemicals are mixed in proper proportion to the incoming water, thus causing precipitation. Mixing may be done by hydraulics or by mechanical equipment:

a) **Baffle mixing:** Acceptable practice for a baffle mixing basin, either around-the-end or over-and-under, is that a velocity of 1.5 feet per second be maintained for the first third of the basin, a velocity of 1.75 feet per second through the last third. In many cases, a flash mechanical mixer at the entrance of the baffled basin will assist in the coagulation process. The flash mixer should have a shaft speed of 350 to 750 revolutions per minute (rpm) and the time of the mix should be from 1 to 5 minutes.

b) **Mechanical mixing:** Mixing should be adequate to disperse the chemicals thoroughly in the raw water prior to its entrance into the flocculation basin. If the mix is done by high-velocity mixers, such as by pump or turbine-type mixer, the velocity should be 5 feet per second or greater, and the retention time should be 1 minute, more or less. If a flash mixer is used, it should be as discussed above under baffle mixing. Paddle-type mixers equipped with variable speed regulators should provide a peripheral speed of 1 to 3 feet per second and a retention time of not less than 5 minutes.

Air mixing devices provide benefits in addition to agitation or mixing of the chemicals. Air being blown through the basin provides this additional treatment in the removal of taste and odors from the water. It also provides oxidation of iron and manganese to aid in their removal. Air mixing units should be designed to provide three stages of agitation: violent, intermediate, and quiescent. The requirement of the air supply is at least 0.5 cubic feet or air per square foot of tank area or 0.05 to 0.20 cubic feet of air per gallon of water. The time of aeration contact of the water should be from 10 to 30 minutes. Air mixing devices may be used in conjunction with either baffle or mechanical mixers.

The final step of mechanical mixing is that of mechanical flocculation, that is, the slow agitation of the fine flox to hold it in suspension so that the size may build up before the sedimentation process. Mechanical flocculators should be of variable speed, with the peripheral speed of the paddles being from 0.5 to 2.0 feet per second. The basin should be sized to provide a retention period of from 20 to 40 minutes.

All conduits carrying coagulated water to the sedimentation basin should be designed to provide a velocity of 0.5 to 1.0 feet per second. Less than 0.3 feet per second, the floc will settle, while at greater than 1.0 feet per second, it will be broken up. Therefore, the 0.5 feet per second is considered the optimum velocity for coagulated water.

4) **Sedimentation basin:** The sedimentation basin is provided for the removal of the floc from the coagulation process. In a well-operated plant, the majority of the purification process takes place in this coagulation process. Therefore, the design of this basin is of importance. The sedimentation basin consists of a diffusion wall and the basin itself. The diffusion wall acts to diffuse the floc over the entire width of the basin. It should be located from 5 to 10 feet from the end of the basin. The size and number of the slots within the wall should be based on a velocity of 0.4 to 0.8 feet per second through the slots and to provide uniform distribution and velocity across the basin. The diffusion wall may be omitted, provided that other methods are used to give this uniformity of distribution.

The sedimentation basin should provide a theoretical detention period of at least four hours. The velocity of flow through the basin should be from 0.3 to a maximum of 1.0 feet per minute. The length-to-width ratio of the basin should be from 2 to 1 to a maximum of 3 to 1. The depth of the basin should be from 10 to 16 feet.

The bottom of the basin should be sloped to a drain sized to empty the basin within a period of 4 hours. An independent overflow should be provided for each basin. The outlet device for settling tanks should be either of the submerged weir-type or large openings to prevent high velocities.

5) **Solids contact or up-flow units:** In fairly recent years, the solids contact or up- flow unit has come into use. This compact unit provides chemical mixing and sedimentation where the settled floc is recirculated to mix with the newly formed floc, thus providing a heavy floc for settling. This type of unit also provides for continuous sludge removal by an automatic timing device. Even though these units have been used in the water softening processes for many years, their application to clarification of turbid waters for community systems requires care in design and operation.

The maximum rise rate of the clarification of water should be in the range of 1.00 to 1.25 gpm per square foot of clarification area. In certain instances where the quality of raw water contains a high concentration of hardness, the rise rate may be increased to 2.25, but in doing so considerable study should be made of the year-round water quality.

The detention period for this type of treatment unit should not be less than 2 hours, with a mixing and flocculating zone averaging not less than 35 minutes. At the greater rise rates, the detention period may be lowered accordingly, but not less than 1 hour. The mixing device should be a variable speed type agitator with a ratio of 2 to 1, and should be designed so that there will be no dead space in the bottom zone of mixing. Sludge removal from the unit should be somewhat continuous, with the sludge removal mechanism controlled by an adjustable automatic timer. The effluent weir of the unit should provide for uniform collection of the clarification area at a lead not to exceed 7 gpm per foot of weir length.

6) **Rapid gravity sand filters:** Sand filters provide the final treatment for the removal of suspended matter for the water. Filtration as a treatment process provides greatest efficiency when a layer of floc creates a mat on the sand surface to filter through. Sand filters are designed on the basis of filtration and backwash rate. The filtration rate is established at 2 gpm per square foot of filter area. In certain instances, this rate may be increased to 3 gpm per square foot of filter area, in which case the plant design and operation must be extremely good that the backwash water rate is established at 24 inches rise per minute. This will give a flow of 15 gpm per square foot of filter area. When Anthrafilt is used, the backwash rate should be around 18 inches rise per minute. To meet these two flow conditions, the filter underdrain and other controls must be designed properly.

For proper operation of a filter, proper controls and gauges must be installed on the filter. These include rate-of-flow controller, loss of head gauges, rate of flow gauges, and wash water rate gauges. Filter bottoms must be adequate to provide filtration and backwash water flows. Bottoms may be either cast-iron manifolds, concrete cast-in-place, or precast plates or blocks. Surface agitator or sweeps are very desirable for proper cleaning of the filter media during the washing cycle.

Filter sand should be of the following specifications:

- Effective size: 0.35 to 0.55 millimeters;
- Uniformity coefficient: 1.70 or less;
- Dust content: Less than 0.5 percent; and
- Depth of sand: 24 to 30 inches.

If Anthrafilt is used, the following specifications apply:

* Effective size: 0.65 to 0.75 millimeters;
* Uniformity coefficient: 1.7 or less; and
* Water wash rate: should be adjusted to give 18-inch rise per minute.

Supporting gravel for the filter media should consist of:

* First 3-inch layer of 5/8- to 1-inch size stone;
* Second 3-inch layer of 3/8- to 5/8-inch size stone;
* Third 3-inch layer of 3/16- to 3/8-inch size stone; and
* Fourth 3-inch layer of number 10 to 3/16-inch stone or Torpedo sand.

To provide for proper expansion of the filter media curing the washing cycle, the minimum distance from the top of the wash through to the top of media should be at least 27 inches. Wash water troughs to the top of the media should be at least 27 inches, and should be of dimensions adequate to carry the maximum wash water rate. The spacing of such troughs should not be greater than 5 to 7 feet.

7) **Chemical feed equipment:** Chemical feed equipment should be of dependable make and accuracy to provide correct dosages. They should be sufficient in number so that split dosage is unnecessary for any one machine except the chlorinator, which is best operated from a panel and has several points of application. Dry chemical feeders should have capacities ranging from 50 to 100 pounds per million gallons of water treated. The upper limit should be provided for the coagulant feed machines.

Chlorinators and chlorine cylinders should be kept in separate rooms from the other feeders for safety reasons. Leakage of chlorine gas will react with dampened metal surfaces to cause corrosion. The chlorinator room should be provided with an exhaust fan located approximately 6 inches above the floor and with a capacity sufficient for two complete air changes per minute. The fan control should be located outside the chlorinator room. A gas mask also should be stored outside the room.

When fluoride compounds are added to the water, the storage of the fluoride chemicals should be kept separate from other chemical storage so as to eliminate the danger of mixing this chemical with other water treatment chemicals.

TOPIC 4: Auxiliary Facilities for Surface Water Treatment

Additional auxiliary facilities are required as part of the water plant:

1) **Laboratory:** The laboratory for proper operation and control should be equipped adequately for daily bacterial and chemical analysis of water quality.

2) **Chemical storage:** Storage areas for treatment of chemicals, should be ample to store at least a month's supply of chemicals. It is preferable to have several months of storage area so that the chemicals may be bought in bulk quantity. All chemical storage areas should be located in the driest portion of the plant to prevent moisture difficulties.

3) **Finished water storage:** This essentially involves clear wells and clear well storage. The basins or reservoirs used for storage of finished water should provide at least 10 to 12 hours of storage based on the maximum consumption expected during that period of time. Together with elevated storage, the total storage should be 1 day's supply, but not less than 75,000 gallons. The clear wells should be constructed of concrete or fabricated of steel. These reservoirs or basins should be covered, and have

watertight access manholes. This should be well-ventilated and contain adequate overflow pipes and drains. All such outlets or openings should be screened with at least a 16-mesh screening. The inlets should be arranged so circulation of water is provided.

4) **Finished water pumping:** Finished water pumping facilities should consist of redundant pumping units. A generator driven by a gasoline or diesel engine should be available as auxiliary power for the plant and pump operation in case of outside utility power failure. For larger installations, pumps should vary in number and capacity from 100 percent to 150 percent to 250 percent of plant capacity.

TOPIC 5: Ground Subsurface Water Sources

Ground water sources or wells usually are characterized by water quality that has a pH from 6.8 to 7.4 (a low concentration of iron, which varies from 0.1 to approximately 1.0 parts per million), and, in some areas, by the presence of hydrogen sulphide. With the low pH, the carbon dioxide concentration increases to create corrosion problems. The range of carbon dioxide extent from about 3 to 5 parts per million at a pH of 7.4 to about 20 to 25 parts per million at a pH of 6.0. With the corrosion problem and the presence of iron and possibly hydrogen sulphide in ground water, the use of certain treatment is sometimes a must with ground water supplies. Treatment facilities involving the removal or prevention of such trouble will be discussed below.

1) **Aeration:** Aeration provides for the exchange of gases so that either removal or oxidation of matter will take place in the exchange interface of water and air. Aerator are used to eliminate carbon dioxide from the water, release hydrogen sulphide, aid in the decomposition of organic matter, control taste and odors, and in the oxidation of iron and manganese for removal of these troublesome compounds. Aerators are usually of the gravity type in which the water is discharged through nozzles into the air and allowed to trickle over cascades or perforated plates or trays with or without special type of media. Other types of aeration processes use the blowing of air through the water media.

   Aerators should be designed so that adequate exposre time and contact may be had. It has been found that coke trays and spray fountains provide around 75 percent removal of carbon dioxide. Hydrogen sulphide is loosely held and is easily eliminated through the gaseous phase exchange of aerators.

2) **Pressure filters:** Pressure filters using sand and gravel have been used in the field of water treatment and swimming pools for many years. This use in the treatment of ground water supplies is also common. Pressure filters using sand and gravel, crushed pyrolusite ore (natural manganese oxide), or ion-exchange resins are in common use in water treatment. When used with aerators and sedimentation basins, they are effective in the removal of iron and manganese, or in conjunction with chemical treatment for the removal of excess hardness from water.

   For removal of iron and manganese, pretreatment in the form of aeration and ph adjustment to above 7.0 is required prior to sedimentation, which is followed by filtration. In this process, the iron and manganese are oxidized by aeration to form a precipitate, which is settled; the filter in this case is used to remove the fine suspended matter that remains in suspension. The oxidation of the iron and manganese takes place as the water flows over the contact media of coke located on trays within the aerator.

   A filter that has been pretreated with sodium permanganate is used to retain the oxidized iron and manganese. Such beds must be cleaned and regenerated by the use of sodium permanganate.

   Pressure filters should be designed with proper piping for backwater, waste times, loss of head gauges, and sight glass indicators. Pressure filters operate at pressures of 20 to 45 pounds per square inch.

Pressure filters may be used in the softening process of water in conjunction with either solid contact units or conventional coagulation treatment units using the lime-soda chemical precipitation process. In such cases, filters must be employed to provide for the removal of suspended matter from the water.

Special types of units using filter boxes with zeolites, in place of filter media, are used for the softening of water. In the ion-exchange process, the ions of magnesium and calcium, components of hardness in water, are exchanged within the ion-exchange resins of sodium ion. In other words, the ions that which cause hardness in water are removed and an ion, which does not possess hardness characteristics, is placed into solution in their place. When the resins become saturated with calcium and magnesium ions, it is then necessary to regenerate the resins by use of salt brines to remove the calcium.

3) **Corrosion control and chlorination:** Other equipment commonly used in the treatment of ground waste supplies, deals with the problem of corrosion control, the pump and distribution system, and the application of chlorination and fluorides to the water in proper dosage. The dosages of these materials are relatively small; therefore, in most cases, they are applied in solution form. All such solution feeders should be of dependable make and of a capacity to provide chemical dosages up to some 10 parts per million, considering the percent of solution strength available and the discharge rate of the source.

Summary Statement

In many instances, it is desirable to use more than one of the surface or ground water sources described in this chapter for a given community water system. The overall design would be based on the development of the individual sources in accordance with what has been discussed previously.

References:

1. Ameen, Joseph S., S.M. Sanitary Engineer. *Community Water Systems. A Source Book.* 5th ed. High Point: Technical Proceedings, 1997.

2. Mays, Larry W. *Water Systems Distribution Systems Handbook.* Department of Environmental and Civil Engineering, Arizona State University. New York: McGraw-Hill, 2000. Endorsed by the American Water Works Association.

CHAPTER 3: Water Quality Standards: A General Perspective on Water Quality

The objective of a community water distribution system that provides drinking water is to deliver sufficient quantities of water where and when it is needed at an acceptable level of quality (the desired chemical and physical characteristics of the water). For the water to be high quality, it must first be free of all harmful bacteria or the index organisms (which will indicate that pathogenic bacteria may find their way into the water supply). (2)

Next, the water must be free of objectionable taste and odors that may be caused by either undesirable chemicals or organisms. It follows that drinking water must be low in concentrations of troublesome minerals, such as iron, sulphur, manganese, calcium, magnesium, and other agents that will make the water unsuitable for use by excessive discoloration or hardness, or nonpotable from the standpoint of high chemical content. Finally, the water must be noncorrosive so that it will not react with plumbing fixtures or pipelines to cause failure of such lines, necessitating frequent replacement, or cause the staining of plumbing fixtures.

The quantity of available supply must be such that the maximum daily demands of the community are satisfied at all times, even during extended periods of drought or after years of community growth. The water supply for most communities not only needs to meet consumer demand but it needs to meet needed fire-flow demand where fire hydrants are installed for the protection of built areas of the community. The same water that supplies domestic taps also supplies fire hydrants, unless there is a separate water system for fire protection.

Although water quality may be acceptable when the water leaves a community treatment plant, transformations can occur as water travels through a distribution system. In the past, water distribution systems were designed and operated mainly on the basis of hydraulic reliability and economics, with little attention paid to the water quality in the distribution system, except when serious problems arose from citizen complaints. This attitude has changed significantly as community water suppliers realize the important influence that time spent in a water distribution system, referenced as resident time, can have on water quality, health concerns, and the conditions of underground water mains.

This chapter is about current water quality standards in the United States including the Safe Drinking Water Act of 2004, factors leading to water quality deterioration in distribution systems, The EPA criteria for "safe" water, and the recognition that there will be a continued "tightening on water quality standards" in the future. All of this is going to require communities to have the methods for monitoring water quality conditions and techniques for modeling water quality transport and transformation. (1)

Factors Leading to Water Quality Deterioration

Water storage facilities, after-water treatment, and the pipe network in a community that transports water from the water storage location(s) to supply point of consumption, constitutes a complex network of entrapped water where uncontrolled chemical and biological reactors can produce significant variations in water quality in both time and space. The primary factors of water quality deterioration in distribution systems include the following:

- contamination via cross-connections or from leaky pipe joints;
- corrosion of iron pipes and dissolution of lead and copper from pipe walls and joints;
- loss of disinfectant residual in storage facilities with long resident time (this can also occur from long resident time in water mains where the flow velocity is inadequate to keep all of the water moving);
- bacterial regrowth and harboring of opportunistic pathogens;
- supply sources going online and offline;
- reactions of disinfectants with organic and inorganic compounds resulting in taste and odor problems;
- increased turbidity caused by particulate resuspension; and
- new formation of disinfection byproducts, some of which could be suspected carcinogens.

The driving factors affecting water quality in a distribution system:

- the quality of the treated water fed to the system;
- the material and condition of the water pipes, distribution system valves, and storage facilities that make up the water system; and
- the amount of time water is retained in the system.

With reference to the last item, it is most important to understand that on a looped pipe network system, the water reaching any particular consumer is actually a blend of water parcels that may originate from different sources at different points in time and follow different flow paths. (See Chapter 4—Topic 5: Designing and Reinforcing Municipal Water Supply Systems Using Looped Water Mains.)

Looped water mains can have an enormous influence on the relation between residence time and water quality. Actions that can be taken to improve water quality or prevent its deterioration in the distribution system include

- changes in treatment practices;
- pipe repair;
- periodic flushing of the water system through fire hydrants;
- relining or replacing pipes; and
- modifications to the water supply operation by circulating water in storage.

Finding an optimal combination of these actions can involve trade-offs between cost, hydraulic reliability, and risk impaired water quality. The issue of water quality for a given community will be referenced in future chapters.

Definitions

The terminology associated with water systems is very specialized and extensive. Throughout all of this text, care is taken to define terms. Important terms frequently used pertaining to water quality issues in water distribution systems are defined below. (2)

Advective Transport: The movement of water quality constituents at the same mean velocity and direction as the bulk carrier fluid.

Aerobic Zone: An area where dissolved oxygen is present.

Alkalinity: Ability of water to neutralize changes in pH.

Anaerobic Zone: An area devoid of dissolved oxygen.

Biofilm: A consortium of microorganisms attached to a solid surface along with a surrounding slimy matrix of extra cellular organic polymers.

Bulk Reactions: Reactions that take place within the volume of water not in contact with the wall of a pipe.

Coliform Bacteria: A group of bacteria associated with the intestinal tract of warm-blooded animals as well as vegetable matter and soil, the presence of which is considered to be an indirect indication of possible fecal contamination.

Disinfection Byproducts: Products of the reaction between a water disinfectant (such as chlorine or ozone) and naturally occurring organic matter in water, some of which are suspected human carcinogens.

Dispersive Transport: Movement of a water-quality constituent caused by concentration gradients.

Euclerian Approach: A modeling framework that assumes a fixed frame of reference when developing conservation of mass, momentum, and energy flow for a fluid control volume.

First-order Reaction: A chemical reaction where the rate of growth or decay of a constituent is proportional to its concentration.

Heterotropic Bacteria: The class of single-cell microorganisms that require organic carbon from both respiration and cell synthesis.

Lagrangain Approach: Modeling framework that assumes a moving frame of reference when developing conservation of mass, momentum, and energy flow for a fluid control volume.

Opportunistic Pathogens: Organisms that may exist as part of the normal body croflora but under certain conditions cause disease in compromised hosts, such as in the elderly, newborns, victims of Acquired Immunodeficiency Syndrom (AIDS), and cancer patients receiving chemotherapy.

Pipe Wall Reactions: Reactions occurring between water quality constituents and materials originating from the wall of a pipe, such as released iron or biofilm slime.

Standpipe: A ground level storage tank whose height is greater than its width.

Tracer Chemical: A nonreactive chemical whose presence is used to track the flow path and travel time of water originating from a particular point of addition.

Topological Sort: A renumbering of the nodes in a directed node-link network, so that all links directed into a given node are connected to nodes of a lower number.

Tubercle: An encrustation growing inward from the wall of a pipe caused by the buildup over time of oxidized corrosion products.

Water Age: The average amount of time that a parcel of water has been in the distribution system.

Regulatory Requirements

In the United States, the baseline regulatory requirements on water quality are set by the EPA. These requirements are enforced by that agency and by the individual States as have been granted primacy by this agency. Individual States may institute more stringent requirements if they deem them to be necessary. Although regulations historically have applied to the point at which water enters the distribution system, there is a trend toward the development of regulations at the point of use.

Shaping the Future of Water Quality and Water Delivery

The Safe Drinking Water Act passed by Congress has set both the tone and the stage for water quality in the United States for years to come. Before addressing the existing national primary drinking water quality regulations, it is important to cover some important perspectives that are shaping the future of community water supplies with a primary focus on maintaining and improving water quality for consumer consumption. Several primary topics are reviewed below:

The Need to Plan Now!

The following quoted statement provided an important perspective on the future of water supplies in the United States, and a brief but extremely important statement on water quality.

> Water utilities across the United States provide an essential service to millions of people every day. In fact, it is a service that is often taken for granted. Water utilities will continue to be challenged by industry trends, such as increasing customer expectations, more stringent regulatory requirements on water quality, increasing water resource demands, and aging infrastructure, while endeavoring to maintain low utility rates. The major challenge over the next decade will be to understand these trends and focus investments of limited resources to create the greatest benefits for both the utility and the consumer in the long run. (1) (Earl Nightingale. The HDR Company, March 2004.)

The HDR Company also provides a prospectus on six topics that address the future of water quality in the United States. A synopsis of each topic is presented below: (1)

1) **Community:** Consumer sentiment and perceptions will drive changes in water utility communications and services. Expectations for water performance have never been higher. Readily available water quality information, together with highly publicized contaminant issues, such as arsenic, have increased consumer scrutiny regarding the quality and safety of public water sources. The Arsenic Rule, which lowers the maximum contaminant level to 10 μg/L, became enforceable in 2006 and affects 41,000 water systems across the country. Of those, about 3,000 are small or very small water systems that have no water treatment infrastructure and limited financial resources. Recognizing this problem, the American Water Works Research Foundation, assisted by HDR developed and demonstrated arsenic removal technologies appropriate for application to small water systems. This involved bench-scale and full-scale studies of 13 different technology variations with help from universities and utilities in Montana, Arizona, and California. With this information, communities are able to address consumer concerns responsibly and confidently. As utilities move forward in this age of bottled water and "treatment at the tap" technology, water supply professionals need to understand the technology and communication tools to meet customer's changing needs and expectations.

2) **Water supply:** Water is considered to be a finite global resource. The U.S. population is expanding at the rate of approximately 1 percent per year, and is projected to exceed 400 million by 2050. Consequently, protection of water resources and innovative applications of reuse treatment technologies will be vital in providing an uninterrupted supply of safe drinking water for future generations. Watershed protection will continue to be an area of intense activity.

As the cost of water treatment continues to escalate to meet drinking water demand and water quality regulations, focusing on improving the quality of raw water sources has to be paramount. The increasing frequency of drought conditions and natural cover fuel fires and forest fires on watershed lands adds to these treatment challenges. As fresh water is exhausted, desalination will become increasingly popular in coastal areas.

The bottom line is—*Easily accessible fresh water supplies are all but gone!* Some creative solutions to these problems will be presented throughout this manual.

3) **Advanced water treatment technology:** As communities plan improvements to treatment processes, advanced drinking water technologies will become a major factor for future regulatory compliance. One emerging regulatory issue is the removal/inactivation of the pathogen Cryptosporidium. Low-pressure membrane technology and ultraviolet irradiation have developed as two key advanced treatment technologies that can be used to assist water systems in eliminating the Cryptosporidium threat to their customers. Reduced costs and increased competition are taking these extremely feasible alternative technologies. The HDR Company is assisting communities such as Kennewick, Washington, in taking advantage of recent advances in submerged membrane technology to expand the capacity of their existing filter basins, while at the same time improving the quality of water to their customers at minimal increased costs.

4) **Water distribution systems:** Water distribution systems are the next frontier facing regulatory pressure relative to water quality. The Total Coliform Rule and the corrosion requirements of the Lead and Copper Rule are only the beginning. As part of the Stage 2 Disinfectants/Disinfection By-Products Rule, systems will conduct an initial distribution evaluation to find sites in the system that have the highest disinfection by-product levels, and new sampling sites will be selected to capture those peak concentrations. Utilities will be paying more attention to the design of distribution infrastructure to reduce water age as defined above. Systems will look for distribution system management practices to minimize potential problems and better control and improve distribution system quality. Of special concern is compliance with both existing and proposed regulations relative to aging distribution systems.

5) **Asset management:** Identifying and prioritizing in investments in operations, maintenance, and capital improvements is a critical challenge facing municipal administrators and managers. This is especially true as water utilities strive to maintain or increase level of service, meet regulatory requirements, and reinvest in an aging infrastructure through improved asset management. Water utilities need *now* to examine their business practices and how decisions are to be made to meet the ever-increasing demands by the EPA for a more stringent delivery of improved drinking water quality.

Understanding the life cycle and the different practices and procedures for supplying water to consumers, and the system needs to supply both maximum consumer consumption demands and fire flow demands simultaneously, is a real challenge for any water system. Typically when normal consumption increases, water supply for fire protection decreases. This is especially true for older water systems. Therefore, municipal officials, water supply officials, and fire service officials need to examine municipal water supplies from a systems approach that considers community assets and the impact that the water system has on fire insurance rates in the city. This is essential for redesigning a water system and optimizing the way the water supply business is done with a focus on minimizing water delivery performance failures.

6) Reuse/Recycle: HDR states, "*...to meet water demands for today and the future, many utilities are developing a balanced portfolio of water resources from traditional and other type of facilities and management techniques.*" (1)

This portfolio needs to include programs such as water transfers between existing systems, improved water conservation, the feasibility of separate water systems for domestic consumption and fire protection, desalination for coastal cities, possible pumping to counties inward from coastal areas, and examining the costs and benefits. Of course, the political considerations of adopting a county-wide water authority to monitor water quality throughout the county and prepare disaster management plans to supply potable water in case of a system interruption are extensive. Alternative water supplies for fire protection where there are no fire hydrants recognized by the ISO is another need. At least one mobile water purification unit should be available in each of the 4,001 listed counties in the United States.

Recycled water is being used in a variety of ways, such as ground water recharge, sea water intrusion barriers, landscape and agricultural irrigation, and industrial and commercial uses. Two examples of wastewater reuse are supplementing surface water that support aquatic life where water supplies are either limited due to poor water quality or limited by availability. Advanced wastewater treatment technologies such as new filtration technologies, membranes, and membrane bioreactors provide innovative solutions to support the substitution of reclaimed water for nonpotable uses. In some cases, reclaimed water is used for indirect potable use via ground water recharge.

Environmental Protection Agency (EPA) Standards for Water Quality—2005

The existing National Primary Drinking Water Regulations are divided into five primary categories with four evaluation areas, plus comments for each category. The following material is presented to provide conceptual understanding of these regulations. **All** municipal officials, including the fire service, need to be aware of the complexity of water quality standards and their individual roles in maintaining these standards.

Due to the amount of material to be presented, two chart sets are provided.

🔹 Chart I: Regulation—Name of Contaminant—Allowable Million Gallons per Liter (unless noted)—Health Effects of Contaminant.

🔹 Chart II: Regulation—Monitoring Requirements—Special Comments.

Chart I

Regulations	Name of Contaminant	Mg/L	Health Effects of Contaminant
Inorganic Chemicals (IOCs)	Antimony	0.006	Increase in blood cholesterol Decrease in blood sugar
	Asbestos	7 MFL	Increased risk of developing benign intestinal polyps
	Barium	2.0	Increase in blood pressure
	Beryllium	0.004	Intestinal lesions
	Cadmium	0.005	Kidney damage
	Chromium (total)	0.1	Allergic dermatitis

Regulations	Name of Contaminant	Mg/L	Health Effects of Contaminant
	Copper	TT (AL=1.3)	Gastrointestinal/Liver/Kidney problems
	Cyanide	0.2	Thyroid/Neurological effects
	Fluoride	4.0	Skeletal problems
	Lead	TT (AL=0.015)	Kidney problems, high blood pressure, delays in physical and mental development in infants and children
	Mercury (inorganic)	0.002	Kidney damage
	Nitrate (as N)	10.0	Methemoglobinemia (blue baby syndrome/diuresis
	Nitrite (as N)	1.0	Methemoglobinemia (blue baby syndrome/diuresis
	Selenium	0.05	Hair or fingernail loss Numbness of fingers or toes Circulatory problems
	Thallium	0.002	Hair loss, changes in blood Kidney/Liver/Intestine problems
	Arsenic Rule		
	Arsenic	0.010	Cancer risk/Cardiovascular and dermal problems
	Radionuclides		
	Combined Radium 226 and 228	5pCi/L	Cancer risk
	Gross Alpha (Excluding Radon and Uranium)	15pCi/L	Cancer risk
	Beta Particles and Photon Emitters	4 mem ede/year	Cancer risk
	Uranium	0.030	Kidney problems and cancer risk
Organic Chemicals	**Synthetic Organic Chemicals (SOCs)**		
	2,3,7,8-TCDD (Dioxin)	1.0000003	Cancer risk/Reproductive system problems
	2,4,5-TP (Silvers)	0.05	Liver problems
	2.4-D	0.07	Adrenal gland/Liver/Kidney problems
	Acrylamide	TT	Cancer risk/nervous system Blood problems
	Alachor	0/002	Cancer risk/Eye/Liver Kidney/Spleen problems
	Atrazine	0.003	Cardiovascular problems Reproductive system problems
	Carbofuran	0.04	Blood/Nervous system Reproductive system problems
	Chlordane	0.002	Cancer risk/Liver Nervous system problems
	Dalapon	0.2	Kidney problems
	Di (2-ethylhexyl) adipate	0.4	Liver/Weight loss Reproductive system problems
	1,2-Dibromo-3 chloropropane (DBCP)	0.002	Cancer risk Reproductive system problems

Regulations	Name of Contaminant	Mg/L	Health Effects of Contaminant
	Di (2-ethylexyl)phthalate	0.006	Cancer risk/Liver Reproductive system problems
	Dinosb	0.007	Reproductive system problems
	Diquat	0.02	Osular problems
	Endothall	0.1	Stomach/Intestinal problems
	Endrin	0.002	Liver problems
	Epichlorohydrin	TT	Cancer risk/Stomach problems
	Ethylene Dibromide (EDB)	0.00005	Cancer risk/Liver/Kidney/Stomach Reproductive system problems
	Glyphosate	0.7	Kidney and reproductive system problems
	Hepatachlor	0.004	Cancer risk/Liver problems
	Hepatachlor Epoxide	0.0002	Cancer risk/Liver problems
	Hexachlorcyclopentadine	0.05	Kidney/Stomach problems
	Lindane	0.002	Kidney/Liver problems
	Metoxychlor	0.04	Reproductive system problems
	Oxamyl (Vydate)	0.2	Nervous system problems
	Benzo[a]0pyrene (PAHs)	0002	Cancer risk Reproductive system problems
	Polychlorinated Biphenyls	0.005	Cancer risk/Thymus gland/ Immune deficiencies/Nervous or reproductive system problems
	Pentachloropenol	0.001	Cancer risk/Liver/Kidney problems
	Picloram	0.5	Liver problems
	Simazine	0.004	Problems with blood
	Toxaphene	0.003	Cancer risk/Liver/Kidney/Thyroid problems
Volatile Organic Chemicals (VOCs)			
	1,1,1-Trichoroethane	0.2	Liver/Circulatory/Nervous system problems
	1,1,2-Trichoroethane	0.005	Kidney/Liver/Immune system problems
	1,1-Dichoroethyhlne	0.007	Liver problems
	1,2,4-Trichlorobenzene	0.07	Adrenal gland problems
	1,2-Dicloroehane	0.005	Cancer risk
	1,2-Dichloropropane	0.005	Cancer risk
	Bezene	0.005	Cancer risk/Anemia/Blood problems
	Carbon Tetrachloride	0.005	Cancer risk/Liver problems
	Chlorbenzene	0.1	Kidney/Liver problems
	Cis 1,2-Dichloroethylene	0.07	Liver problems
	Dichlormethane	0.005	Cancer risk/Liver problems
	Ethylbenzene	0.7	Kidney/Liver problems
	Ortho-Dichlorobenzine	0.6	Kidney/Liver/Circulatory problems
	Para-Dichlorobenzene	0.075	Kidney/Liver/Spleen/

Regulations	Name of Contaminant	Mg/L	Health Effects of Contaminant
	Circulatory problems		
	Styrene	0.1	Kidney/Liver/Circulatory problems
	Tetrachloroethylene		
	(PCE)	0.005	Cancer risk/Liver problems
	Toluene	1.0	Kidney/Liver and
	nervous system problems		
	Trans-1,2-Dichloroethylene	0.1	Liver problems
	Trichloroethylene	0.005	Cancer risk/Liver problems
	Vinyl Chloride	0.002	Cancer risk
	Xylenes (total)	10.0	Nervous system problems

Microbiological Contaminants

	Total Coliform Rule	See Comments	Coliforms are bacteria that are naturally present in the environment and are used as an indicator that other, potentially harmful, bacteria may be present. The presence of fecal coliform or E-Coli may indicate potential contamination that can cause diarrhea, cramps, nausea, headaches, or other symptoms.
	Surface Water Treatment Rule (TCR)		
	Turbidity	TT	None. Interferes with disinfection.
	Giardia lambia	TT	Gastrointestinal illness, Giardiasis
	Enteric Viruses	TT	Gastrointestinal and other viral infections
	Ligionella	TT	Legionnaire's disease
	Heterotrophic Plate Count	TT	None. Used to measure a variety of common bacteria in water. Lower bacteria concentrations indicates better maintained water systems.
	Interim Enhanced Surface Water Treatment Rule (IESWTR)		
	Turbidity	TT	None. Interferes with disinfection.
	Cyptosporidium	TT	Gastrointestinal illness, Cyptosporidiosis
	Long-Term (1) Enhanced Surface Water Treatment Rule (LTIESWTR)		
	Turbidity	TT	None. Interferes with disinfection.
	Cyptosporidium	TT	Gastrointestinal illness, Cyptosporidiosis

Disinfectants and Disinfection Byproducts
Stage 1 D/DBPR Disinfectants

	Chlorine	4.0 as Cl (2)	Eye/Nose irritation Stomach discomfort
	Chloramines	4.0 as Cl (2)	Eye/Nose irritation Stomach discomfort/Anemia
	Chlorine Dioxide	0.8 as ClO (2)	Anemia/Nervous system problems

Regulations	Name of Contaminant	Mg/L	Health Effects of Contaminant
	Disinfection Byproducts		
	Total Triholomethanes	0.080	Cancer risk Potential reproductive system effects
	Liver/Kidney		
	Haloacetic Acids	0.060	Cancer risk
	Chlorite	1.0	Anemia/Nervous system problems
	Bromate	0.010	Cancer risk
	Total Organic Carbon	TT	

Chart II

Regulations	Name of Contaminant	Monitoring Requirements
Inorganic Chemicals (IOCs)	Antimony	Surface water—once a year Ground water—once every 3 years for IOC's, not including Asbestos, Lead, Copper, Nitrite or radionuclides
	Asbestos	Once every 9 years
	Copper	Same as Lead
	Lead	Followup monitoring every 6 months after corrosion controls are initiated or optimized. Systems consistently meeting AL can reduce monitoring to once per year and then every 3 years after that.
	Nitrate (as N)	Ground water—annually Surface water—quarterly initially, then annually Repeat frequently as determined by State requirements.
	Arsenic	Same as IOCs.
	Combined Radium	Sample point is the distribution system entry point that is representative of all sources being used. Four consecutive quarterly samples must be taken at all sample points.
	Beta Particles and Photon Emitters	The primary agency must designate the system as vulnerable. Once deemed vulnerable, quarterly samples must be taken for beta emitters and annual samples taken for Tritium and Strontium-90 at the distribution system entry point. Compliance is based on the running annual average of four quarterly samples taken at each sample point.
	Uranium	Same as Combined Radium and Gross Alpha
Microbiological Contaminants	Total Coliforms	For both surface and ground waters, the total number and location of samples is based on the population served and a system-specific sampling plan.
	Surface Water	Grab samples at least every 4 hours for continuous monitoring. Continuous chlorine residual required for systems greater than 3,300 persons. One to four grab samples per day are allowed for systems with less than 3,300 persons.
	Treatment Rule (SWTR)	

Regulations	Name of Contaminant	Monitoring Requirements
	Turbidity	Conventional and direct filtration systems must measure combined effluent turbidity at least every 4 hours and continuously monitor turbidity of each individual filter.
	Filter Back	No monitoring is required, but water systems must collect and retain the following information for review by the primary agency:

- Copy of recycle notification and information submitted to the primary agency.
- List of all recycle flows and frequency returned.
- Average and maximum filter backwash flow rate (gpm) along with the average and maximum filter backwash duration.
- Typical filter run time and written summary of how the filter run length is determined.
- The treatment provided for recycle flow.
- Design data on the treatment provided for the recycle flow, dimensions for recycle flow treatment processes, hydraulic loading rates, type of chemicals used along with the dose and frequency used, and frequency at which solids are removed.

Wash Rule

Long-Term Surface Water Rule (LTSWR)

	Turbidity	Conventional and direct filtration systems must measure combined filter and effluent turbidity at least every 4 hours and continuously monitor turbidity of each individual filter.

Disinfectants and Disinfection Byproducts

	Disinfectants	Monitor the same as for the Total Coliform Rule. Compliance is based on running annual arithmetic average of monthly averages. Daily sample at distribution system entry point.
	Disinfectant	Four quarterly distribution samples. Compliance on running annual average of quarterly average.
	Chlorine	Daily sample at distribution system entry point.
	Bromate	One sample per month (ozone systems only) and running annual average.
	Total Organic Carbon Byproducts	Source treated Water TOC sample once a month.

The Future of Environmental Protection Agency Regulation

The following information pertains to the *projected regulatory schedule* that covers a time period to 2011. It is imperative that local governments keep current on the actual implementation or modifications to these proposed regulations.

1) Regulation on the contaminant *Alicarb*. This applies to a contaminant that could adversely affect the nervous system. Regulations originally were proposed in 1991. EPA will re-examine risk assessment and occurrence data and determine what further action is appropriate. Final promulgation is set with a projected effective date of 2008.

2) Radon Rule Adjustment. Radon presents a potential cancer risk. The proposed rule adjustment would be to increase the monitoring requirements. A proposed effective date for new regulations is set for 2007.

3) Revisions to the Total Coliform Rule (TCR). Coliforms may be indicators of harmful bacteria in the water supply. The revisions proposed address requirements to finished water quality in the distribution system as well as to evaluate alternative monitoring strategies. A firm proposal is scheduled for final promulgation in 2007, with an effective date of 2010.

4) A proposed Stage 2 regulation on disinfectants and disinfection. This has implications for health problems associated with water supplies, including cancer risk, potential reproductive system effect, liver, kidney, and nervous system problems. The rule would apply to all community and nontransient noncommunity water systems that add a disinfectant other than UV or deliver water that has been disinfected. Final promulgation is set with an effective date of 2008.

5) Long-Term 2: Enhanced Surface Treatment Rule applies to all public water systems that use surface or ground water. It is most important to protect against gastrointestinal illnesses. Beginning 6 months after the actual promulgation, monthly monitoring of the water source for Cryptosporidium for 2 years will be to characterize the source water. For small systems, E-Coli concentrations will be in lieu of Gryptosporidium. Based upon the conditions found, the levels of additional removal beyond that provided by conventional treatment or direct filtration will be identified. To meet the additional removal requirements, the water system would select the appropriate tool(s) from the Microbial Toolbox. Note that ultraviolet light disinfection would be allowed under this rule. For implementation, large and medium systems where the population served was greater than 10,000, have 3 years after source water characteristics or 6 years after rule promulgation to meet the higher removal requirements. Small systems will be on a delayed schedule.

6) Ground Water Rule (GWR): The new ground water rule also is aimed at preventing Gastrointestinal disorders and illness. It specifies monthly sampling and testing of the source water for E-Coli, Enterococci, or Coliphage for systems not treating to a 4-log inactivation or removal of viruses and raw water from a sensitive aquifer. Continuous monitoring of the treatment techniques for water systems serving a population greater than 3,300, while smaller water systems will be required only to take grab samples. The effective date for this rule is 2007.

7) Six-Year Review of Regulations: In June 2002, the EPA proposed to review the total Colifom Rule (TCR), based on its review of the existing regulations. However, based on revised risk assessments, several contaminations could be proposed "out of cycle." Potential "out of cycle" revisions are possible for Atrazine, TCE, Fluoride, Chromium, and particularly Chromium VI.

8) Contaminant Candidate List (CCL) and Regulatory Determinations: The CCL is a starting point to make a regulatory determination on whether to rule a specific contaminant or not. EPA proposed the first regulatory determinations in April 2002, and proposed **not** to regulate any of the priority contaminants from the first CCL. This assessment on some contaminants such as Perchlorate and MTBE, are being completed and are likely to be the first "out of cycle" regulations from CCL regulatory determinations. No effective date has been set at this writing.

9) Endocrine disruptors: Some scientists have proposed that certain chemicals might be disrupting the endocrine system of humans and wildlife. To that end, the EPA is developing a screening program to identify the chemicals that should be monitored to determine the spread of endocrine disruptors. Criteria to select priority chemicals for screening are to be finalized in the near future. The EPA plans to propose a rule to implement the screen process. There is no established timeline on this subject.

Water Quality Monitoring

It is difficult to use monitoring data alone to understand the fate of transformation of substances in drinking water as the water moves through the distribution system. Even medium-sized cities have thousands of miles of pipes, making it impossible to achieve widespread monitoring. The flow paths and the travel paths of water through these systems are highly variable because of the single-point feed or the looped layout of the pipe network and the continuous changes in water usage over time. The common use of storage facilities in the system makes things even more variable. At different times of the day, a specific location in the pipe network might be relatively new water from the treatment works when storage tanks are being refilled, or old water when storage tanks are being emptied. It is usually impractical to experiment on the entire distribution system by seeing how changes in pumping schedules, storage facility operations, or treatment methods affect the quality of water received by the consumer.

For these reasons, mathematical modeling of water quality behavior in distribution systems has become an attractive supplement to monitoring. These models offer a cost-effective way to study the spatial and temporal variation of a number of water quality constituents including:

- the fraction of water originating from a particular source;
- the age or time of residence, of water in the system;
- the concentration of a nonreactive tracer compound either added or removed from the water system (e.g., fluoride or sodium);
- the concentration and loss rate of secondary disinfectant (e.g., chlorine); and
- the concentration and growth rate of disinfection byproducts such as THMs, and the number and mass of attached and free-flowing bacteria in the system.

The models can be used to assist managers to perform a variety of water quality related studies. Examples include the following:

- calibrating and testing hydraulic models of the system through the use of chemical tracers;
- locating and sizing storage facilities and modifying system operations to reduce the age of the water;
- modifying the design and operation of the system to provide a desired blend of waters from different sources;
- finding the best combination of pipe replacement, pipe relining, pipe cleaning, reduction in storage holding time, location, and injection rate at booster stations to maintain desired disinfection levels throughout the system;
- assessing and minimizing the risk of consumer exposure to disinfectant byproducts; and
- assessing the systems vulnerability to incidents of external contamination.

References:

1. HDR Company, Washington Office, 8403 Colesville Road, Suite 910, Silver Spring, MD 20910-3313.
2. Mays, Larry W. "Water Distribution Systems Handbook." Ch. 9 in *Water Quality* by Walter M. Grayman, Consulting Engineer, Lewis R. Rossman, U.S. Environmental Protection Agency, and Edwin E. Geldreich, Consulting Microbiologist. New York: McGraw-Hill, 2000.

CHAPTER 4: Water Distribution System Design Concepts

Water Supply Source Classifications

Chapters 2 and 3 both described two basic sources of water that are used for community water supply systems as follows:

1) Ground water sources (wells), which indicate that the water is below ground level or below the earth's surface. It is necessary to dig or bore a well in order to tap into the water source. This is also generally referred to as a "well-water source." Artesian wells may have water that is trapped below a rock formation so that the water is under pressure. When an auger breaks through the rock formation, there may be sufficient water under pressure to force the water to the earth's surface where it is then pumped to the water treatment facility and either to finished water storage or pumped directly into the distribution system. All other well sites require a well pump to raise water to the earth surface. Depending on the depth of the well, the pump may be able to transport the water directly to the water treatment facility; otherwise, one or more pumps are needed in series to move water to the treatment location. **Figure 4-1** shows a typical arrangement of a well water pump to treatment facility arrangement.

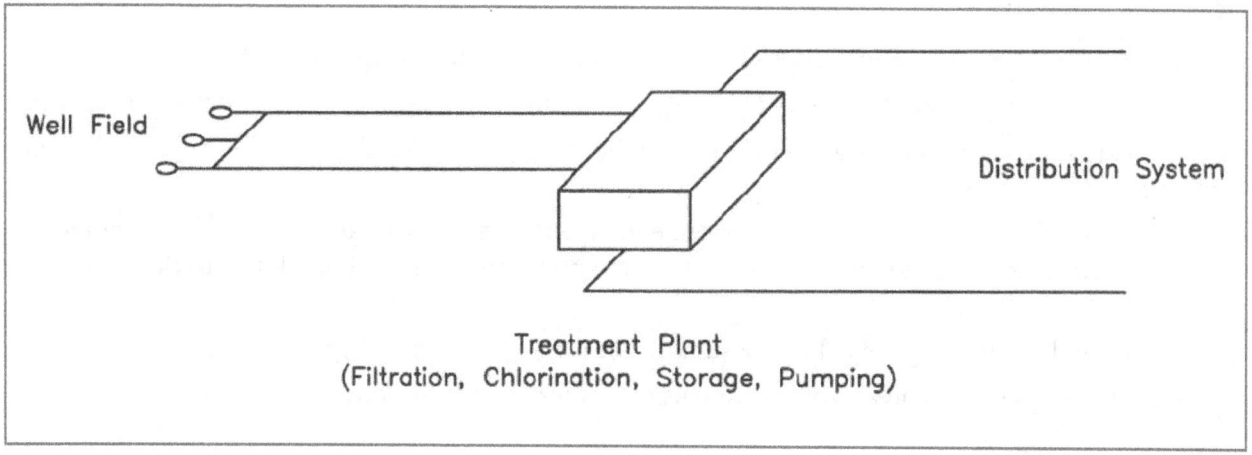

Well Field

Distribution System

Treatment Plant
(Filtration, Chlorination, Storage, Pumping)

Figure 4-1

2) Surface waters represent the second general classification of water supplies. This means that the water is taken from aboveground water sources that include both large and small natural lakes, and high and low reservoir lakes formed by manmade dams to retain the water, rivers, and streams.

Water System Components

It is essential to understand the basic types of water supply systems and the physical arrangement of system components prior to discussing either design or evaluation concepts associated with these components. There are two basic types of water supply systems to create water pressure within the distribution system to supply water to built areas of a community, and to provide required water pressure to fire hydrants located on the water supply system. These two types of water systems that supply water under pressure for consumer consumption and fire protection are

1) **Gravity feed systems.**

2) **Pumping pressure systems:** Each of these systems must take water from a supply source, pass the water through a treatment plant, and then transport the water into the distribution system. It needs to be recognized that community water systems can be divided into four basic classification according to the water source:

   a) High or low reservoirs that hold nonpotable water for gravity feed.

   b) Pumping station systems that use ground water from streams, rivers, canals, man-made or natural lakes, and other special provisions for impound water. In these types of systems, the raw water is pumped from the source point to the treatment plant and then either pumped directly into the distribution system or into storage to be used on demand by the community.

   c) Pumps at well sites that pump water to the treatment facility. Based on the difference in elevation between the treatment facility and the community to be served, the water may flow by gravity through the distribution systems, or there may be the need for another pumping station.

   d) A combination of gravity flow and one or more pumping stations to transport the water from the source point to all of the water demand points on the distribution system.

Water System Classifications

Communities up to about 15,000 population may have a relatively simple, straightforward classification for their municipal water supply. However, large cities can have a very complex water system. There are some basic water system classifications that have been established to provide a basic reference for examining the adequacy and reliability or water systems. These are identified as follows along with referenced illustrations:

1) **High-level reservoir system:** This classification refers to a water source that is at least 100 feet in elevation above the treatment facility in order to provide sufficient head pressure so that nonpumping station is required. The water should be a natural impounded lake with a properly installed flow pipe between a source point in the lake and the treatment facility, or the impounded lake may be created by a dam on a stream or a runoff area in a valley that will allow the water to back up behind the dam. This is illustrated in **Figure 4-2**.

   This type of system gives the opportunity to develop a very reliable water supply system. If there is sufficient elevation difference between the water intake source on the reservoir system and the distribution piping in the community, it is possible to design a water system that does not require one or more pumping stations. The head pressure for supplying water to the treatment plant and from the treatment plant to the distribution system is sufficient to meet both consumer demand and needed fire flows. This is also a very economical system since there is no substantial power requirement to run the water system beyond normal lighting requirements and requirements for establishing the level of water quality needed for the community.

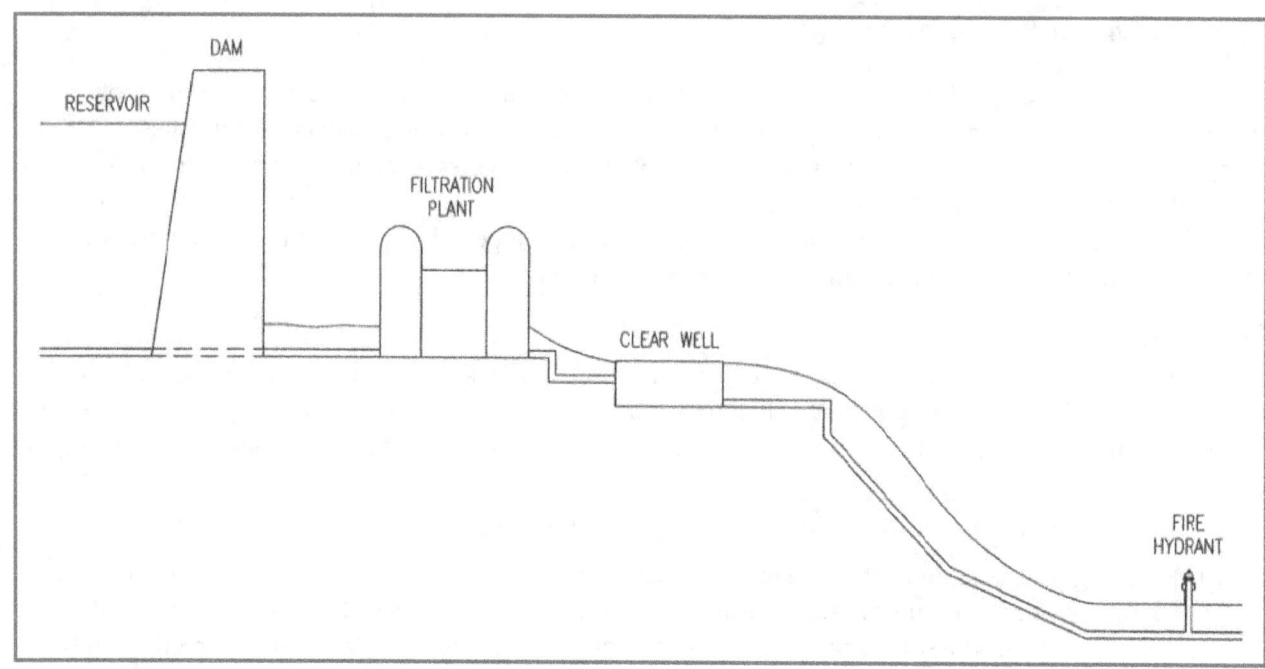

Figure 4-2

High-level reservoir systems are found in high hills and mountain regions of the country and around the Great Lakes, where the lake plain has foothills of 200 to 500 feet above the lake level. These systems are highly reliable under all weather conditions that may cause disruption to the other types of systems discussed below. Typically, reservoir water supplies can be controlled at the surface level so as to minimize the amount of water treatment necessary to meet current and future water quality requirements.

A gravity flow system that does not require a pump interface is considered the most reliable type of water system, since there is no mechanical component to break down or fail when the power source goes down. For the same reasons, it is the most economical type of water supply system to operate.

2) **Low level reservoir systems:** Low-level dams, usually up to about 20 feet high, are typically built on steam flow that is considered reliable year round. The reservoir basin may be bulldozed to form a small lake that will supply water to communities of populations up to about 25,000. One type of low-level reservoir provides a channel from a major river or lake to form a small lake behind a low-level dam. If there is an expected variation in the water level from the primary source, such as a drought or heavy freezing in the winter months, a gated pipe with an earth cover is used to prevent backfeed to the primary water supply source. A basic configuration for a low-level dam is depicted in **Figure 4-3**.

Low-level retention dams typically require a pumping station to transport water to the treatment plant and, if the land area is relatively flat, a second pumping station to pump treated water directly to the distribution system or to elevated storage to provide the required pressure and volume to meet instantaneous flow demand. The elevated storage can be designed to minimize the direct pumping requirements.

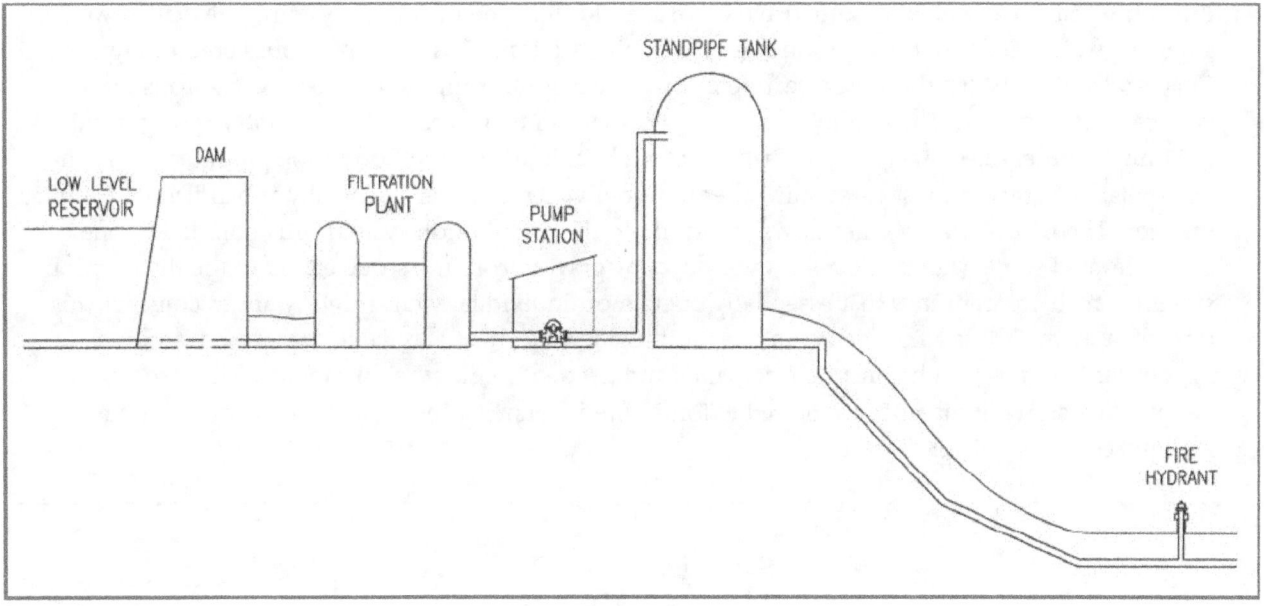

Figure 4-3

3) **Direct pumping systems:** **Figure 4-4** illustrates how a direct pumping station feeds water to the treatment plant and then a second pumping system transports water to a storage holding area, such as a clear well, to a standpipe storage tank that is maintained full as domestic consumption varies throughout a single day. This minimizes the time the pump or pumps actually have to run. The pumps also may be designed and arranged to pump the treated water directly into the distribution system when there is a high demand on the water system. This could occur when there is a major fire in the community.

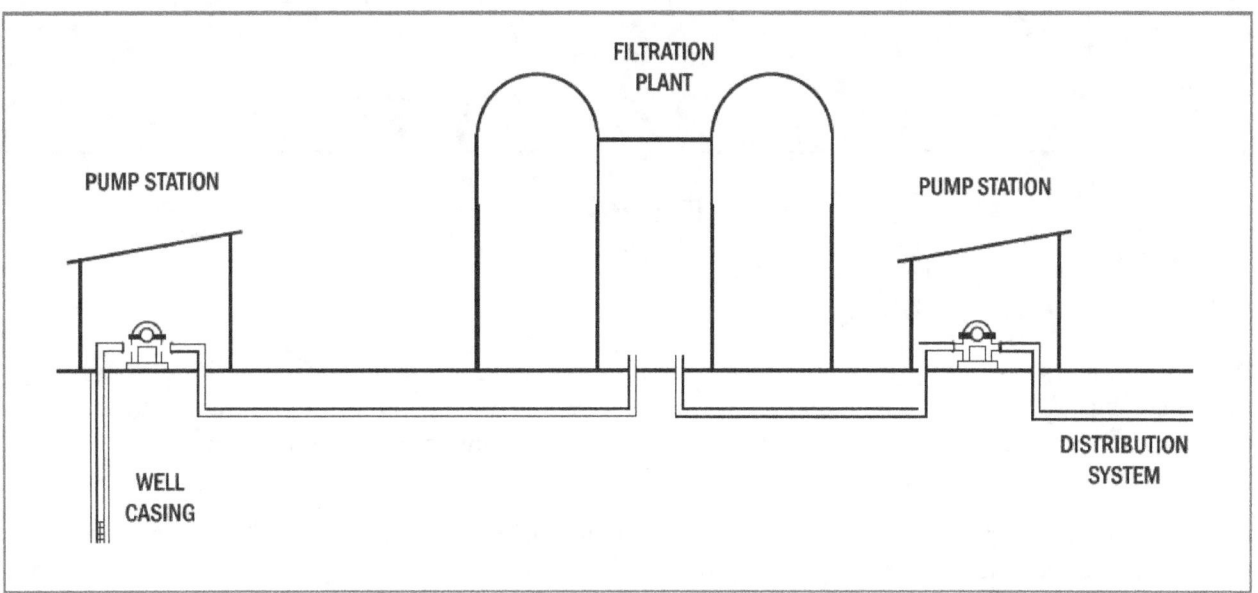

Figure 4-4

4) Pumping station at well sites and gravity storage: In this type of supply system, one drilled well, or a field of wells, feed water to a ground-level pumping station. This concept is presented in **Figure 4-5**. Chapter 3 indicates that the water treatment for ground water supplies may be less rigorous than for surface water supplies. The quality of the ground water in many areas of the country is so good that the only treatment necessary is chlorination through an injection method in the pipes that carry the non-potable water. In most cases, any other required water treatment generally is handled in a similar manner. The treated water either flows by gravity to the distribution system or is pumped to one or more elevated storage tanks. Potable water flows by gravity from the storage tank to the distribution system. Small communities that essentially operate off individual wells might want to consider this type of system. This is a more efficient way to provide water that is reliable year round, and is a system that can meet both consumer demands and needed fire flow requirements. This concept should be investigated with reference to the reduction in fire insurance premiums as the result of having a recognized water delivery system.

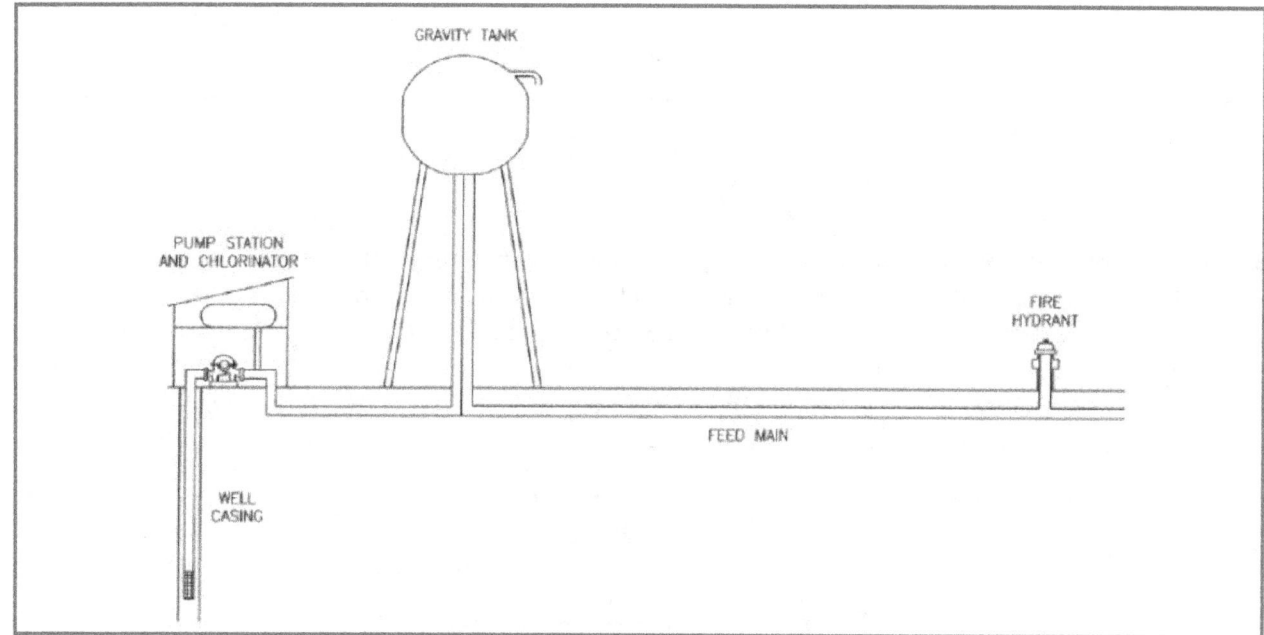

Figure 4-5

Composite Water Supply Systems

While the descriptions and configurations reviewed above cover the basic types of community water supply systems, specific conditions of areas served by the water system and, most importantly, the topography, may require a "composite system" that uses components from more than one of the "typical water systems." Some examples:

💧 Adding pumping stations to a gravity reservoir system to increase pressure and volume during peak demand periods including a fire flow requirement

💧 Gravity tank supplies often have more than one storage tank to meet daily consumption requirements. On smaller water systems, gravity tanks may be needed to meet both volume and pressure requirements in case of a "working structure fire."

- Booster pumping stations may be installed where there is a need for more than one service level based on pressure demand.

- Well fields may have a number of different lift pumps that transport water to a filtration plant instead of localized purification.

- Direct pumping into the distribution system may be supplemented by gravity tanks that "float" on the system to maintain pressure and flow characteristics during different demand periods through the day and night.

- Gravity tanks are especially useful for improving the reliability of any water system.

System Demand and Water Design and Flow Criteria

Overview

The following information is a synopsis of the subject title by the AWWA *Manual of Water Supply Practices* M-31. (7) This material reviews the *significant* relationships between the various demands placed on a water distribution system and outlines recommended design-flow criteria. However, before system demands or design-flow criteria can be reviewed, a basic understanding of a water distribution system and its specific components is needed. The following is a basic review of water distribution systems and the requirements placed on specific elements of a water system.

Methods of water distribution: Water is dispersed throughout the distribution system in a number of different ways, depending on local conditions or on regulations and requirements that influence water system design. The common methods of water distribution to the pipe network are reviewed under the following titles.

1) **Gravity distribution:** This is possible when the treated water source is a retention pond, clear well, or storage tank at some needed elevation above the community. In this type of system, sufficient pressure is available due to gravity to maintain water pressure in the mains for domestic consumption and fire service demand. This is the most reliable method of distribution if the piping leading from the treated water source to the community is adequate in size and safeguarded against accidental breaks. Higher pressures for firefighting, however, requires the use of mobile fire department pumpers and, in some cases, stationary booster pumps on the water system to provide needed fire flows at representative fire hydrants with a residual pressure of 20 psi.

2) **Pumps and elevated storage:** Through the use of pumps and elevated storage, the excess water pumped during periods of low consumption is stored in elevated tanks or reservoirs. During periods of high consumption, the stored water supplements the water that is being pumped. This method allows fairly uniform flow rates and pressures throughout the water system. Consequently, this method generally is economical because the pumps may be operated at their rated capacity. Since the stored water supplements the supply used for fires and system breakdowns, this method of operation is fairly reliable. However, it is necessary that fire department pumpers be available to boost the pressure from fire hydrant to delivery water through hose streams at the proper nozzle pressure to confine, control, and extinguish developing structural fires and other related fire events. A more complete understanding of these concepts is presented under **distribution storage** below.

3) **Pumps without storage:** When stationary pumps are used to distribute water, and no storage is provided on the distribution system, the pumps force water at the required volume and pressure directly into the mains. The outlet for the water is through domestic taps on the system or through fire

hydrants. This is the least desirable type of distribution system because a power failure could interrupt the water supply. In addition, as consumption varies, the pressure in the water mains is most likely to fluctuate. To conform to varying rates, several pumps are made available to add water output when needed, a procedure requiring constant attention at the water plant. Another disadvantage is the fact that the peak power demand of the water plant is likely to occur during periods of high electric power consumption, thus increasing power costs to operate the water system. However, one advantage of direct pumping is that a large stationary fire pump may be used on demand for structure fires. This pump increases the residual pressure to any desired amount permitted by the construction of the water mains.

Rates of Water Use

Three historical or predicted water demand rates are involved in the discussion of water system demand and design flow rate criteria for both consumer consumption and needed fire flow. These are as follows: (7)

1) **Average daily consumption:** This is the average of the total amount of water used each day during a 1-year period (usually expressed in million gallons per day, MGD)

2) **Maximum daily consumption:** This is the maximum total amount of water used during any 24-hour period in a 3-year period. This number should consider and exclude any unusual and excessive identified used of water that would affect the calculation. Such abnormal uses would include a water main break, a large-scale fire, or an abnormal industrial demand. This is often referred to as the MDC rate.

3) **Maximum hourly demand:** This is the maximum amount of water used in any single hour, of any day, in a 3-year period. It is normally expressed in gallons per day. It is determined in gallons per day by multiplying the peak hours by 24. This can also be express as MHD.

Distribution System Appurtenances

Piping and valve arrangement: *Water systems typically have three classifications of pipe used to transport to demand points throughout a community. These are identified as follows:*

1) **Primary feeders:** These are large pipes, usually with diameters ranging from 12 to 36 inches, based on the size of the population served. Primary feeders transport water form the water treatment plant to corporation line of the community and/or to major water storage locations within the community.

2) **Secondary feeders:** These are connected to the primary feeders to transport water along the major streets of the community. Secondary feeders need to be in place to supply all commercial property, public buildings, and private sector buildings that have a needed fire flow over 1,000 gpm. Secondary feeders typically are 10 to 16 inches in diameter.

3) **Distributor mains:** These are used to transport water from the secondary feeders to individual streets in the areas of the community that have small businesses like convenience stores and gas stations but, more importantly, along residential streets. The minimum pipe size should be 6 inches and, based on the system design, a possible dead- end pipe may need to be 8 or even 10 inches.

The sizes of pipe associated with the three classifications of pipe in a typical water system are approximations. The needed pipe size throughout the built-upon areas of a community is based on the hydraulic gradient of the community, consumer consumption profiles through the community, needed fire flow at representative locations throughout the community, and, quite importantly, the two methods for laying and connecting pipe throughout the community. The traditional pipe system design is referred to as a *Single-Point Feed System*. This is illustrated in **Figure 4-6**. In this case, water moves from the treatment plant to the community corporation line with a single primary feeder. The primary feed, in turn, supplies the secondary feeders along the main streets of the community, and the distributor mains supply the block frontage along the residential streets. Note that any demand point on the system for either consumer consumption or fire flow through a fire hydrant, is fed from one direction only. In the single- point feed system, the pipe sizes need to meet the maximum daily consumption demand plus the needed fire flow. This results in a larger pipe than is needed under normal daily usage, without any fires. The second major weakness of this type of system is that there may be a lot of dead-end mains in the residential areas and at the end of secondary mains. This leads to the stagnation of water which rapidly reduces the quality of water, as discussed in Chapter 3.

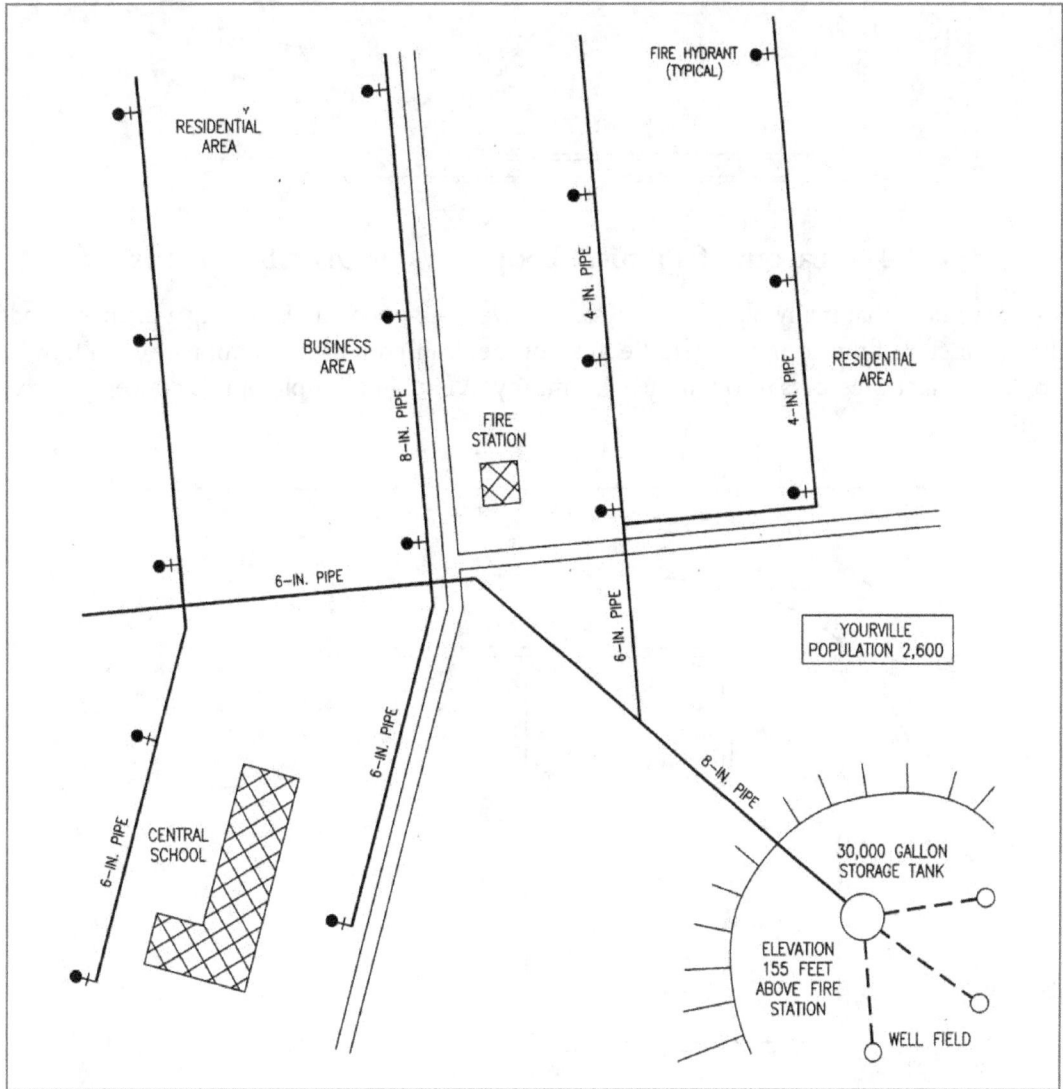

Figure 4-6: Features of a Small Community Water Distribution System

The more modern approach to water system design is to loop all the water mains, or cross-tie the mains, so that at any demand point the water is supplied from two directions. A basic *Pipe Looped* water system is depicted in **Figure 4-7**. This allows the designing engineer to develop a hydraulic model of the system and to determine mathematically the proper size of pipe according to flow paths to meet the consumer and needed fire flow demand points.

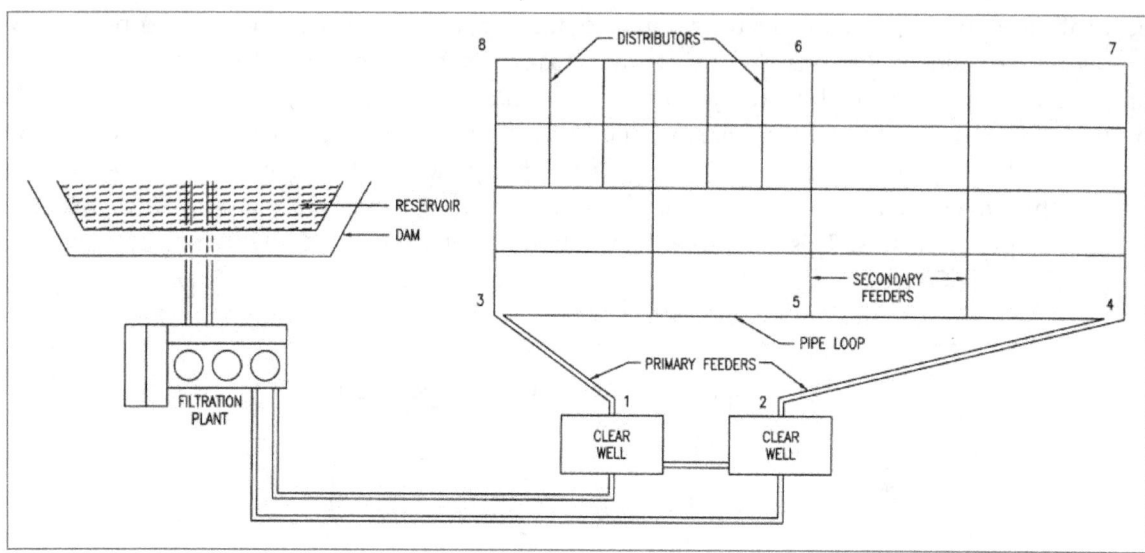

Figure 4-7: Layout of a Typical Looped Water Distribution System

It is important to note that many older water systems have been updated. By laying a primary feeder around the perimeter of the community to tie in all of the dead-end mains to improve both flow distribution and water pressures through the community. A simple example of this concept is presented in **Figure 4-8**.

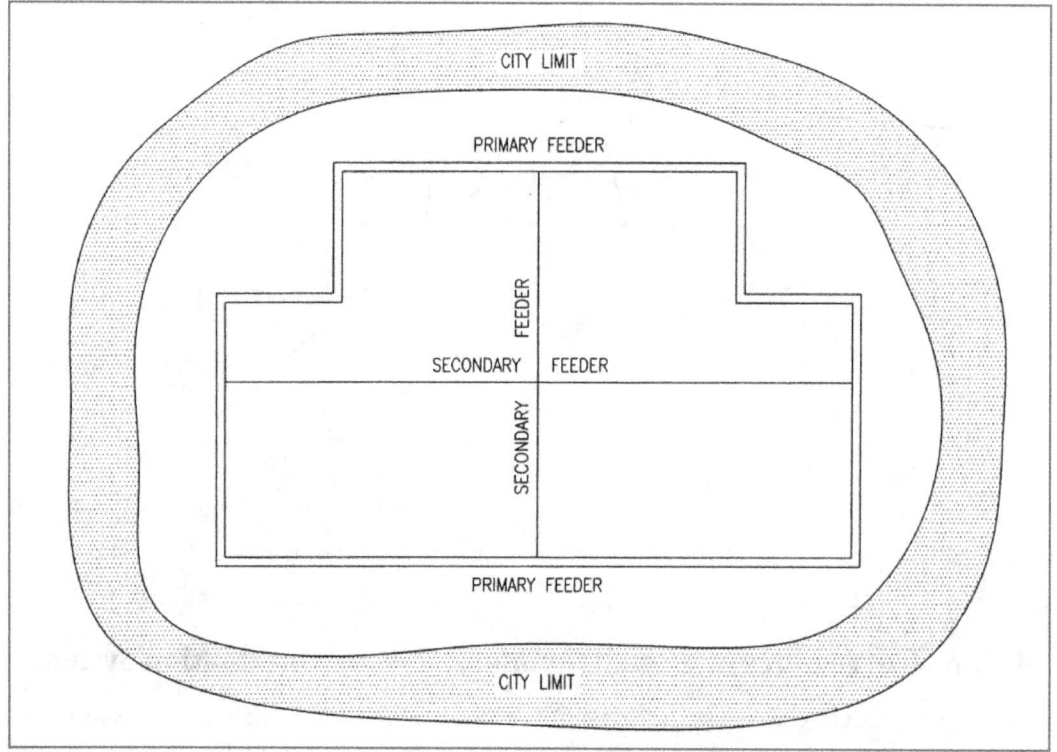

Figure 4-8: Typical Small City Distribution System

The following provisions conform to the current recommended practices of the AWWA.

⬥ In small cities, the primary feeders should form a loop about 3,000 feet in length or two-thirds of the distance from the center of the community to the outskirts. The primary feeders should have control valves not over 1 mile apart, and the mains connecting them also should be valved where they connect. This configuration is necessary so that interruptions in service will not require shutting down the feeder main.

⬥ In large cities, the primary feeders should be arranged into several interlocking loops, with the mains not over 3,000 feet apart. Looping allows continuous service through the rest of the primary mains even when one portion is shut down temporarily for repairs. Under normal conditions, looping also allows supply from two directions for large fire flows. Large feeders and long feeders should be equipped with blow-off valves at low points and air-relief valves at high points.

⬥ The secondary feeders carry large quantities of water from the primary feeders to points in the system in order to provide for normal supply and fire fighting. They form smaller loops within the loops of the primary mains by running from one primary feeder to another. Secondary feeders should be spaced only a few blocks apart. The spacing allows concentration of large amounts of water for fire fighting without excessive heat loss and resulting low pressure.

⬥ Small distribution mains for a grid over the area to be served supply water to fire hydrants and service pipes for residences and other smaller buildings. The size of these mains usually will be determined by the design flow. In residential areas, particularly where there are heavy uses for lawn watering, it may be necessary to determine the maximum consumer water demand.

This network of pipes throughout all the outlying sections of a community may consist of single mains or dual mains as depicted in **Figures 4-9** and **4-10**. **Table 4-1** presents a comparison of the repair procedures needed for single and dual main systems.

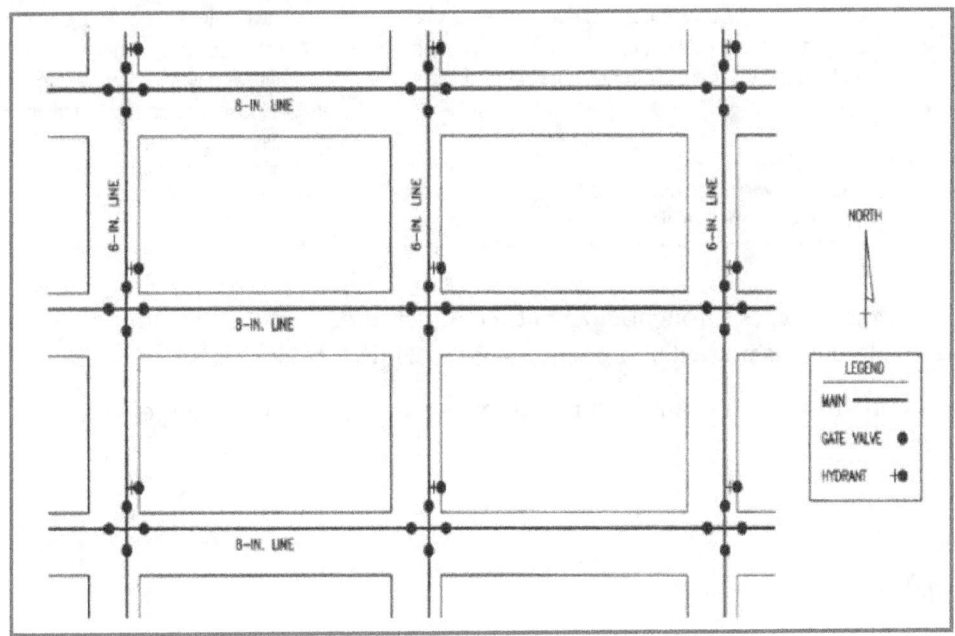

Figure 4-9: Single Main System Layout

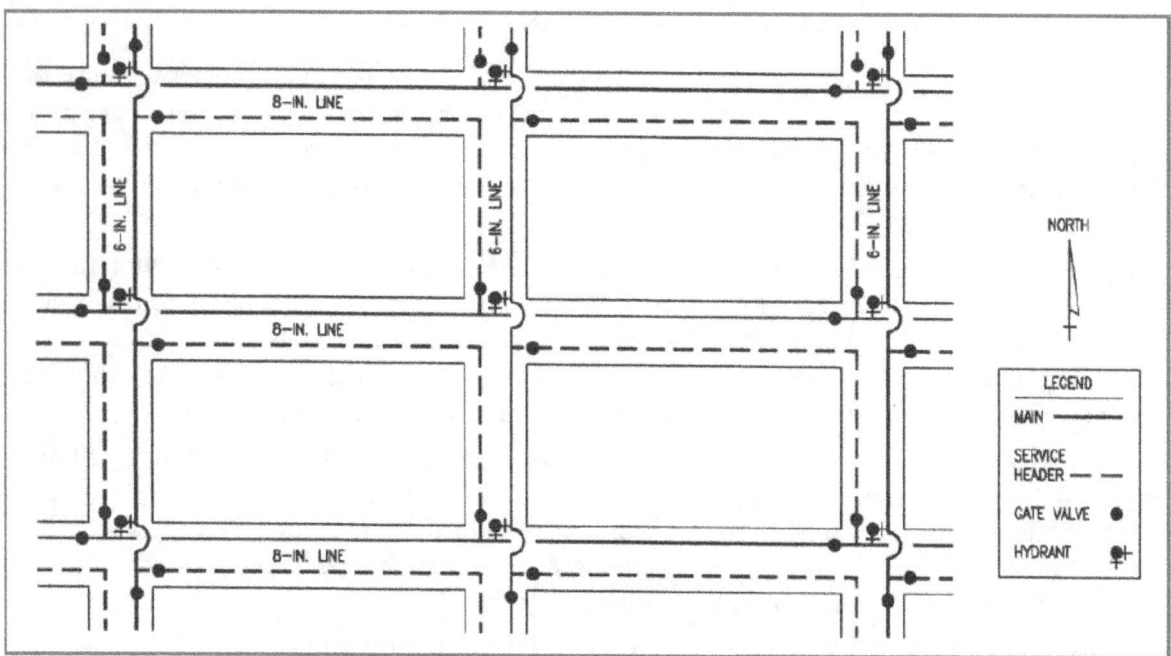

Figure 4-10: Dual Main System Layout

Table 4-1: Comparisons of Repair Procedures for Single and Dual Main Water Systems

Breakdown Points	Single Main Systems	Dual Main Systems
Fire hydrant	Close on hydrant valve.	Close to hydrant valves.
East-west mains	Close four valves on main. This dead-ends supply to one hydrant and cuts out intermediate service taps. It also dead-ends supply to eastern intersections.	Close two valves on main and one valve on each of the connected hydrants and service headers. Hydrants and service headers remain in commission. House service taps on mains are placed out of commission. Service headers are dead-ended.
North-south mains	Close two valves on main. One hydrant and intermediate house taps are placed out of commission.	Same as east-west main

The described network of pipes throughout all but the outlying sections of a community may consist of single mains or dual mains as presented in **Figure 4-9** and **Figure 4-10**.

The AWWA generally installs valves on a community water system on mains in the following recommended manner:

◆ four at crosses;

◆ three at tees; and

◆ one on each hydrant branch lines.

This is done in order to facilitate sectionalizing the water system for repairs and cleaning of the water units. In dual main systems, service headers are added on the south and west side of the streets, and piping is generally placed between the street curb and the sidewalks. In a scheme suggested by Ballou, valves are installed in the following manner: (2)

♦ one on each main at intervals of two blocks;

♦ one at the intersection of the service header; and

♦ one on each end of hydrant branches that are more than 10 feet long.

Service operations and conditions in breakdowns and cleaning can then be compared, as was shown in **Table 4-1**.

The hydraulic advantages of dual main systems over single main systems depend more or less on local conditions, because service headers generally do not contribute to flow outside their own area. More importantly, if broken, piping in dual main systems will not impair the usefulness of hydrants or create dead-end mains.

The dual main systems frequently will compare favorably in cost with single main systems. The advantages are that repairs can be made and new services laid without interfering with traffic or damaging the pavement, repairs can be made at lower costs, leakage usually is reduced, since service lines (which are liable to have heavy leakage) are reduced in length, and service pipes need not be laid where the main passes vacant lots, thereby reducing idle investment and minimizing resident time of water in the mains. Minimum use service mains are the common source of serious water system leaks.

Distribution piping should be sized to meet design flow as determined hydraulic analysis on water system flow gradients. The minimum size of water mains, for providing fire protection and serving fire hydrants, is 6 inches in diameter. Larger size mains will be necessary to achieve required fire flow and maintain residual pressure specified for both domestic consumption of fire flow. The piping must meet the minimum standards specified in **Table 4-2**.

Table 4-2: Minimum Standards for Distribution Piping

Appurtenance	Minimum Standard
Pipe Lines:	
Smallest pipe lines in the network	6 inches
Smallest branching pipes that are dead ends	8 inches
Largest spacing of a 6 or 8 inch grid	600 feet
Smallest pipes in the high value district	8 inches
Smallest pipes on principal streets in the central business or commercial district or business property complexes	12 inches
Largest spacing of primary and secondary feeders	3,000 feet
Valves:	
Spacing in single and dual main systems	
Largest spacing on long branches	800 feet
Largest spacing in commercial and high value districts	500 feet

Hydrant Locations

All built-upon areas of a community should be served by a water distribution system that not only provides taps for consumer consumption, but also provides approved fire hydrants for installation at locations and with spacing considerations for convenient use by fire department pumping equipment and to meet needed fire flows in the proximity of the buildings to be protected. In North America there are two basic methods for the distribution of fire hydrants in a community. The first method is commonly used in Canada. It is mentioned briefly here because the method has been used in some of the northern States that border Canada. This method is fully described in *Water Supply for Public Fire Protection, A Guide to Recommended Practices.* (8)

The second method is a linear method document in the ISO's *Fire Suppression Rating Schedule* that is used to establish Public Protection Classifications (PPCs) using a scale of 1 to 10, with 1 representing the best possible classification and a 10 indicating no recognized fire protection for establishing insurance rates. While the classification system is somewhat complex, the lower or better the Class number, the lower the property fire insurance rates. In other words, a community classed as an 8 would be expected to pay much higher fire insurance premiums than a community that has a Class 4 rating. The community water supply evaluation accounts for 40 percent of the entire evaluation. The distribution and location of fire hydrants, based on needed fire flows for commercial buildings, is an important part of this evaluation. (4)

The needed location of fire hydrants in a community according to the Grading Schedule criteria can be briefly summarized as follows: The ISO evaluation procedures examines a number of representative locations throughout a community based on the population of the community; the number of installed fire hydrants on the water system; and property types, which include commercial, industrial, and residential properties along with educational facilities, public buildings, such as courthouses, libraries, hospitals, and all other places of public assembly. Buildings upon land areas within a community that do not have a fire hydrant within 1,000 feet of a structure also are evaluated. For a fire hydrant to be credited in the community survey, it must be within 1,000 feet of the property to be protected. Flow tests are conducted to determine that each fire hydrant delivers a minimum of 250 gpm at 20 psi residual pressure for a duration of 2 hours. Installed fire hydrants that do not have this capability receive no credit for insurance rating purposes although they may be used for limited fire suppression capability.

Of importance is that the fire hydrant spacing in a community is also evaluated. For individual property evaluations, the following maximum gpm flow is limited by the hydrant spacing from the risk site. Measurement is made according to the distance that a fire hoseline has to be laid to the fire risk building.

- 1,000 gpm may be credited for a fire hydrant within 300 feet.
- 670 gpm for fire hydrants from 301 to 600 feet.
- 250 gpm for fire hydrants from 601 to 1,000 feet.

Additional spacing criteria may be adopted by local building codes and/or the governmental authority having jurisdiction over a specific building. Commercial buildings that are protected by approved automatic sprinkler systems may have special requirements by individual insurance companies for fire hydrant locations that may be used to reinforce the water supply volume and pressure by fire department pumpers to control and extinguish interior fires. Municipal officials need to call these special needs to the attention of community property protected by automatic sprinkler systems.

In addition to the above provisions, it is a recommended practice that the maximum lineal distance between fire hydrants along streets in congested areas and high fire risk areas with frame buildings and/or high combustible storage (such as lumber), be located 300 feet apart and a maximum of 600 feet, in light residential areas with building separations of **over** 50 feet. Other good practices for the installation of fire hydrants calls for at least one fire hydrant at every street intersection, in the middle of long blocks (especially where the needed fire flow exceeds 1,300 gpm), and near the end of long dead-end streams.

Hydrants should be required within large complexes that are accessible to fire department apparatus equipped with mobile pumps.

A key consideration: It is essential that the planning of fire hydrant locations be a **cooperative** *effort between the community water department, the fire department, the building and zoning department, and with the insurance carrier for large commercial and industrial complexes. The proper location of the fire hydrants definitely can have a positive impact on property fire insurance rates.*

Installation of Fire Hydrants

The proper installation of fire hydrants on municipal water systems has to give consideration the construction features of the fire hydrant. (10) These include, but are not limited to the following features for providing adequate and reliable water supplies for fire protection.

- The nominal diameter of the bottom valve opening needs to be at least 4 inches for supplying two 2-1/2-inch outlets. However, it is recommended that at least one large diameter outlet be provided for the connection hose to a mobile fire department pumper. Today, it is generally recommended that the bottom valve be a minimum of 6 inches.

- The net area of the hydrant barrel and foot piece at the smallest part is not to be less then 120 percent of that of the net opening of the main valve.

- A sufficient waterway through the fire hydrant needs to be provided to minimize friction loss in the fire hydrant. The hydrant designed should not permit more that 5 psi loss from the main valve intake to the discharge side of the fire hydrant with design flows of no less then 1000 gallons per minute. This information is obtained from the Underwriters Laboratories, Inc. (UL) listing.

- A positive-operating, corrosion-resistant drain or drip valve is to be provided.

- A uniform sized pentagonal operating nut measuring 1-1/2 inch from point to the flat at the base and 2-7/16 inches at the top. The faces should be tapered uniformly and not less than 1 inch.

Fire hydrant bonnets, barrels, and foot piece are generally made of cast iron with internal working parts of bronze. Valve facings should be of a suitable, yielding material such as rubber or a composition material. Fire hydrants are available a number of different configurations, which should be adapted to the installation location.

Types of Fire Hydrants

Two types of fire hydrants are generally in use today. The most common is the base valve or dry barrel, in which the assembly controlling the water supply from municipal water system pipes is below the frost line between the foot valve or "piece" and the barrel of the fire hydrant. This common type of fire hydrant is illustrated in **Figure 4-11.** The barrel on this type of fire hydrant is normally dry, with water being admitted only when there is a fire or when the hydrant is flow tested; other uses of this fire hydrant are discouraged. A drain valve at the base of the barrel is open when the main valve is closed, thus allowing residual water in the barrel of the hydrant to drain out. This type of fire hydrant needs to be installed whenever there is a chance the temperature will go below freezing, because the valve assembly and water supply are installed below the frost line determined from climatic conditions.

The second type of basic fire hydrant is the wet-barrel type which is normally limited to the southern and western States where protracted freezing is most unlikely; temperatures inside the hydrant barrel must remain above freezing at all times. This type of fire hydrant usually has a compression valve at each outlet, but they may have another valve in the bonnet that controls the water flow to each of the hose outlets. This type of fire hydrant is illustrated in **Figure 4-12.**

In the side Figure, a base valve, or a dry barrel is illustrated with nomenclature identified. When installed, the valve is below the frost line. This type of fire hydrant also is known as a "frost-proof" fire hydrant. (Source: *Mueller Company*)

Installation Positioning

The following guidelines have been prepared by the International Fire Service Training Association. (11)

1) The large-diameter outlet on a fire hydrant, normally 4 inches to 6 inches, should be positioned perpendicular to the curb line on streets or to the edge of the roadway where a fire department can connect hose from the fire hydrant to the intake of the mobile fire pump.

2) The maximum connection distance between the large-diameter outlet on a fire hydrant and the intake connection on a mobile pumper should not exceed 15 feet; less than 10 feet is preferable except for special situations such as parking lots and parking garages.

   The responsible fire department should be consulted on special fire hydrant installation problems.

3) There should be no obstructions within 10 feet of any installed fire hydrant. Such obstructions generally include traffic standards, sign posts, utility poles, trees, shrubbery, and fences.

   A *wet-barrel* or California-type fire hydrant maybe used where freezing temperatures are not encountered. There is a compression valve at each outlet. (Sources: Mueller Company)

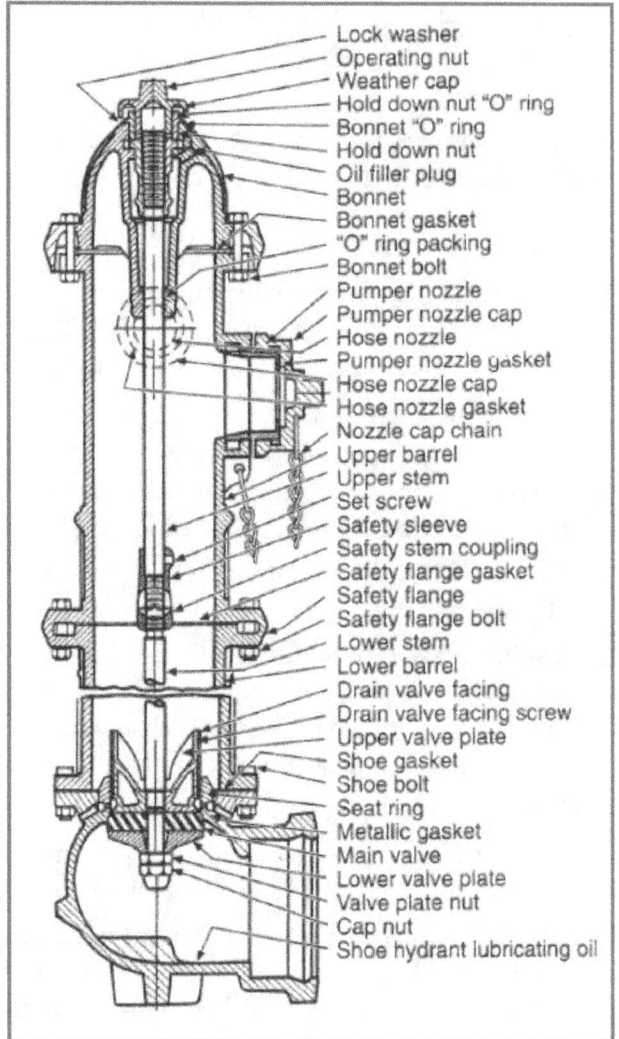

Figure 4-11:
Dry Barrel Fire Hydrant

4) The clearance between the bottom of any hydrant capped outlet to the ground level or hard surface levels is not to be less than 18 inches. To provide for connecting different types of hose fittings to the fire hydrant.

5) All fire hydrants needs to be painted a bright color and that color maintained for visibility by approaching fire apparatus from all directions.

Installation Drainage

Drainage is necessary for fire hydrant equipped with drain ports, and can be provided by excavating a pit about 2 feet in diameter and 2 feet below the base of the fire hydrant and filling it compactly with course gravel or stones placed around the bowl of the fire hydrant to a level 6 inches above the waste opening. If the drip valve of the hydrant is below ground water level, it maybe plugged to exclude ground water. In this case, water should be pumped out to prevent freezing. This same procedure should be followed for older installed fire hydrants not equipped with drain features and that have been install for many years. (10)

General Maintenance of Fire Hydrants

A major item of periodic maintenance recommended on a quarterly schedule is a check for any external damage to each fire hydrant and a check for leaks in the fire hydrant mechanism that include 1) the main valve assembly when the fire hydrant is closed, 2) the drip valve when the main valve is open but outlet are capped, and 3) the mains the supply each fire hydrant. Stethoscope-like listening devices are available to make these checks.

Maintenance routines provide for an operating test, repair of any leaks, and pumping out of fire hydrants where necessary. Threads of the outlet, caps, and the valve steam should be lubricated with graphite. Hydrants should be kept painted on a scheduled basis for proper location during an emergency, but care should be taken to avoid accumulation of paint that might prevent easy removal of caps or operation of valve stems. (10)

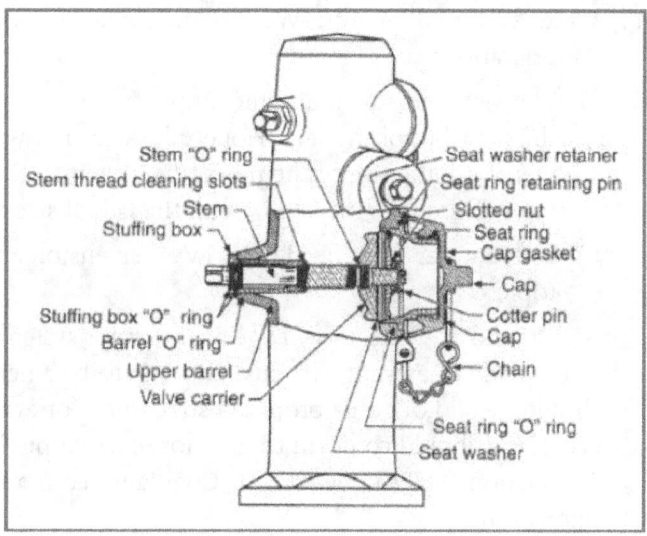

Figure 4-12: Wet Barrel Fire Hydrant

System Evaluation and Design

The designing and evaluating of community water supply distribution systems has to consider the amount of water for the commercial interests, governmental property, educational facilities, and all classifications of residential property as presented above in a general relationship to average and maximum daily consumption demand. At any time of the day, the day of the week, or the week of a given year, a structure fire or other fire emergency such as transportation vehicle fires or, in some cases, natural cover fires may erupt. Water is the primary agent of choice to confine, control, and extinguish structural fires. Some new development in fire extinguishing agents may be used for rapid knockdown of a fire, but a well-developed structure fire still requires established needed fire flows from fire hydrants to control and extinguish developing fires. Each community needs to evaluate and design or modify the design of the community water system to meet present-day needs to address future demands based on growth of the built area and population increases, along with the need to meet EPA criteria for water quality, as discussed in Chapter 3. This will be an ever-increasing demand and challenge for *every* community water distribution system.

Some specific guidelines on consumer consumption requirements and needed fire flows, are established by the ISO, which represents in excess of 130 property and casualty underwriters in the United States in developing advisory insurance rates. The following topics address some fundamental information on understanding 1) water system demands, 2) determining design flow, and the very important topic of 3) water storage on a community water system. This should provide community leaders, municipal officials, fire department officials, water supply superintendents, and consulting engineers on water systems, a common knowledge base so that they all can sit at the same table and have a meaningful dialog about the present and future state of a specific water system and even how it may relate to adjacent water supplies in nearby community water systems.

1) **Water system demands:** Water demands need to be assessed on the basis of the following considerations:

   a) The year and date the water supply system was commissioned or started supplying water through the distribution system. Hopefully, there is a water supply system map that will detail where pipes were laid, the size of the pipes, the pipe material, the location of the valves, the location, size and type of fire hydrants along with the lateral size and valving arrangements.

   b) These water maps need to show all extension and changes to the water system covering the same topics as above.

   A clear and accurate knowledge of the water system in place is needed before discussing changes to be made to existing systems. If this is to be a new community water system, all of these details should be laid out on a proposed street map for evaluation. If an Engineering Firm uses different criteria, it should conform to the most current publications of the AWWA, the National Fire Protection Association (NFPA), the ISO, the Civil Engineering Handbook, and any special State or county regulations.

   There should be no oversight in considering both short-term and long-term goals. The primary objective is to make sure that the community is being serviced adequately. If there are deficiencies in meeting current or future goals because of economic constraints, this needs to be identified for the areas of the community where there may be inadequate flows to meet consumer consumption during peak water demand, so that constraints such as watering lawns and washing cars, can be placed on water usage. If available fire flows do not meet needed fire flows in specific districts of the community, the fire department needs to know these conditions on virtually a real-time basis. The local fire department may need to plan on relaying water from larger supply mains to fire sites using large-diameter hose or the existing water supply may need to be augmented by an alternative water supply using mobile takers from adjacent fire departments under automatic-aid and mutual-aid arrangements. Another alternative is to provide retention ponds in the community to capture runoff. Retention ponds may be outfitted with dry hydrants as a supplementary water supply for fire protection.

   [Sidebar: Refer to Chapter 6 for further discussion of Alternative Water Supplies.]

2) **Planning water demand changes:** Special planning is needed when new water demands will be placed on a specific community water system. This point cannot be overstressed. The amount of construction performed and the amount of construction that realistically can be accomplished to provide adequate service are dependent on when the construction will be needed. Ultimately, final development should be consistent with the utility's ability to provide consumer consumption and fire protection at the same time. Fire protection must not lag behind supplying domestic taps, as often occurs in new residential areas of communities. The planning and installation phase should assure that water supply for fire protection is *never* interrupted. There have been too many large-loss structure fires because the water system was shut down in the vicinity of the fire site.

   The ideal way to develop a water distribution system would be to construct a distribution network of pipe that would adequately serve the short-range and long-range development of the service area. Individual construction projects, developments, subdivisions, and industrial complexes then could be developed without checking for adjustments to ensure that the original design plans remain adequate for all projected consumer consumption and fire protection demand. However, in reality, the best of plans needs to be adaptable to change measures where growth, moves, and demand may decline in older portions of a community. Therefore, the design for the source, through the treatment plant to the distribution system, must provide for growth and change in the delivery demand points. The best water system is the one that is designed with a vision for the future. Existing water systems have to be evaluated and redesigned with a future perspective that includes a rehabilitation and/or replacement of existing system components due to the age factor. The maintenance of old infrastructures may be more expensive than the replacement with a better designed system that will meet both the future needs of water quality and distribution system demand.

The bottom line is that a water supply system cannot remain constant. It is the responsibility of elected officials, water supply superintendents and their staff, hydrologist, geologists, professional civil engineers, rural and urban planners as appropriate, fire officials and fire protection engineers, and representatives from the insurance industry to sit at the table and plan water systems for the future with due consideration to all of the regulations and requirements that are being placed on water systems at this time and in the future, especially as programmed by the EPA. The cost to do this is not going to be any small thing, so the financial planning is just as important as the physical planning.

It is now recommended that every 3 to 5 years, as a minimum, existing water distribution systems be evaluated thoroughly for requirements that would be placed on it by development and reconstruction for a 20-year period into the future. A plan then should be developed for meeting those needs. In this way, individual improvements and projects can be evaluated and made to conform, generally, to long-term development and contingency plans for such events as serious system interruptions caused by natural disasters and terrorism attacks without undue additional expense to either the developer or the utility.

Basic Concepts in Determining Design Flow at System Demand Points

The first step in evaluating or designing a water delivery system is to determine design flows for all of the representative built-upon regions of the community and the community expansion plans. The fundamentals of calculating water system design flow are as follows:

1) Determine the average daily demand.

   [Sidebar: See Chapter 5 for guidance on determining flow demands by areas of the community.]

2) Determine the maximum daily demand or estimate it from the average daily demand.

3) Determine the maximum hourly demand from consumption records or for new areas of the community, estimate from the average daily rate.

4) Determine the ISO needed fire flow that is discussed in Chapter 5. Actually, two methods of determining fire flow are presented for evaluation but only the ISO method is used for establishing PPCs advisory insurance rates used by most property and casualty underwriters.

5) Determine the required flow for sprinkler system requirements plus a supplemental allowance for fire streams if a commercial property is protected by an automatic sprinkler system.

6) Select an appropriate design flow for the community water system. This is the greater of the following two conditions: either the sum of the required fire flow for the most stringent situation which, not considering property protected by automatic sprinklers, plus the maximum daily consumption demand; or the maximum hourly demand, whichever is greater.

The community water supply planning group must exercise good judgment based on the facts at this point. The following example makes this point very clear. It relates to communities that have manufacturing plants or seasonal agricultural processing and canning plants.

In this example, the property in question is identified as a manufacturing facility that uses large amounts of water for only 8 hours during one shift per day, 5 days a week. With relatively little domestic use, the maximum hourly demand would be in excess of what the water supply committee would expect from normal projection data and examining the average daily demand. In this case, the water system serving the facility should be designed to provide the required flow plus the maximum hourly flow. It would be costly and essentially irresponsible to consider treating a fire at this facility on the basis of an average daily demand that has no real meaning in the particular context. Furthermore, commercial properties that have a high peak demand for short periods of time should work with the community to provide private water storage to meet these peak periods. This could save major water treatment costs and pipe design costs to transport water to these specific demand points that have an infrequent need for the supply.

It is very important to design and evaluate a given water system using the design flow as generally determined by the above considerations so as to ensure the adequacy and reliability of the system. Therefore, it is essential that all of the issues relating to water system adequacy and reliability as presented under **Water Supply Concepts and Evaluation Methods** be considered carefully when considering the future design of a new or existing community water system.

Distribution System Storage

Finished water placed in storage within the framework of a community water system, may just be one of the most important aspects of achieving optimal performance, reliability, and cost effectiveness in water system design. It should be understood that water storage facilities have a far-reaching effect on a water systems ability to provide needed fire flows when the fire cannot be immediately extinguished by the first arriving fire company. The two common methods of storing water on a water distribution system are discussed below under:

- ground level storage; and
- elevated storage.

There are relative merits to both types of storage. Prior to examining these two common types of water storage, preliminary considerations need to be examined on the basic purpose of water storage.

Functions of water system distribution system storage: Storage capacity within the framework of a distribution system enables the system to process water at a time when a treatment facility would be unable to meet the systems demands or when the treatment facility is idle. Stored water provides the possibility to distribute the treated or finished water to one or more locations in the service areas that provides the best flow and pressure to the end user.

1) **Advantages:** The principle advantages of water distribution storage include the fact that storage equalized demands on supply sources, water treatment facilities, plus the transmission and distribution mans. As a result, the sizes or capacities of these elements need not be as large as would be required if storage was not provided. Additionally, system flows and pressure can be improved and balanced or stabilized to better serve both consumer and required fire flows throughout the service area. Finally, reserve supplies are provided in the distribution system for other emergencies such as power outages and breaks in the water distribution system.

2) **Storage as a function of meeting system demands and required fire flows:** The location, capacity, and evaluation, if required, of distribution system storage are considerations that are closely associated with system demands and the variation in demands that occur throughout the day in different parts of the system. System demand can be determined only after careful analysis of an entire distribution system by professional engineers in cooperation with local authorities and the water supply planning committee as identified above. However, there are some general rules that may serve as a guide to such an analysis. **Table 4-3** lists daily and hourly variation for a typical city and the resulting treated water storage depletion.

Table 4-3: Water Use and Storage Depletion for Maximum Day in a Typical City

Hour	Ratio of Hourly Demand Rate to Maximum Day Demand Rate*	Hourly Variation in Distribution Storage Reserve Mil Gallons	Cumulative Storage Depletion Mil Gallons
7–8 a.m.	1.00	-0.00	0.00
8–9	1.10	-0.10	0.10
9–10	1.25	-0.25	1.35
10–11	1.28	-0.28	0.63
11–12	1.20	-0.20	0.83
12–1 p.m.	1.18	-0.18	1.01
1–2	1.16	-0.16	1.17
2–3	1.10	-0.10	1.27
3–4	1.00	-0.00	1.27
4–5	1.08	-0.08	1.35
5–6	1.15	-0.15	1.50
6–7	1.30	-0.30	1.80
7–8	1.60	-0.60	2.40
8–9	1.40	-0.40	2.80
9–10	1.25	-0.25	3.05#
10–11	0.90	+0.10	2.95
11–12	0.85	+0.15	2.80
12–1 a.m.	0.70	+0.30	2.50
1–2	0.60	+0.40	2.10
2–3	0.50	+0.50	1.60
3–4	0.50	+0.50	1.10
4–5	0.50	+0.50	0.60
5–6	0.60	+0.40	0.40
6–7	0.80	+0.20	0.20

* Average day—16 million gallons; Maximum day—25 million gallons; constant hourly supply rate (at maximum day demand rate), 24 million gallons per day or 1 million gallons per hour.

Maximum storage depletion.

Of course, each water system has its own requirements. Furthermore, it should be recognized that only in rare cases can distribution storage be justified economically in an amount greater than what will take care of normal daily variation and provide the needed reserve for fire protection and minor emergencies. In systems of moderate size, the amount of water storage available for equalizing water production is 30 percent to 40 percent of the total storage available for water pressure equalization purposes and emergency water supply reserves. In normal system operations, some water from storage should be used each day not only to maintain uniformity in production and pumping, but also to ensure circulation of the stored water for purposes of maintaining water quality. The best economical arrangement for the storage of water for delivery to the distribution system is to bring all storage to full capacity at night when there is minimal domestic consumption and then increase as needed when storage capacity drops to between 40 percent and 50 percent at each storage point location during the day. In the event of a working structural fire in a community at any time of day or night, pump capacity should be brought on line to compensate for the fire flow being used.

Normally, it is more advantageous to provide several smaller storage units in different parts of the system than it is to provide equivalent capacity at a central location. Small pipelines are required to serve decentralized storage, and other things being equal, a lower flow-line evaluation and pumping head result.

Elevated and Ground Storage

Storage within the distribution system is normally provided in one of two ways:

1) Elevated storage.
2) Ground-level storage with high service pumping.

It should be recognized that elevated storage provides the best, most reliable, and most useful form of water storage, particularly for fire suppression requirements.

Ground-level storage: Since water kept in ground storage is not under pressure, it must be delivered to the water demand points by pumping equipment. This arrangement especially limits water system effectiveness for fire protection in three ways:

1) There must be sufficient excess pumping capacity to deliver the peak consumer consumption demand use as well as any fire demand which requires unnecessary investment in pumping capacity, **or** if the pumping capacity is deficient to meet this demand, it could adversely impact on the ISO PPC which drives fire insurance rates in the community.
2) Standby power sources and standby pumping systems must be maintained at **all times** because the system cannot function without the pumps.
3) The distribution lines to all points in the distribution system must be significantly oversized to handle peak demand use, plus required fire flow, no matter where a fire could occur in the community.

However, in hilly areas it is frequently possible to install ground-level reservoirs at a sufficient elevation so that the water would "float" on the distribution system. This may eliminate the need for pumps at the ground-storage facility or reserve pumps only for high demand periods. If the desired overflow elevation can be achieved on a hill, a considerably larger storage capacity can be installed when compared to an elevated tank. This may result in placement of the storage facility on a hill in a less desirable location and then use transmission mains down to the primary feeder pipes. Such placement would provide larger storage capacity than could be achieved by an elevated tank(s) or it would provide the equivalent storage more economically.

Where ground storage is used in areas with high fire risk, buildings with flows of 1,000 gpm or more, the energy would be needed to deliver the water that is lost on the initial delivery of water to the water system. The water must be repumped and repressurized with the consequence of additional or more standby generators and more standby pumps to provide fire flow adequacy. In addition, the system's high-service pumps must be either variable speed or controlled by discharge valves to maintain constant system pressures. This equipment is expensive, uses additional electric power, and requires extensive operation and maintenance. Frequently, the additional capital costs for pumps, generators, and back-up systems, and the long-term energy costs significantly increase the costs of a ground storage system.

Elevated storage: Properly sized elevated water tanks provide dedicated fire storage and are used to maintain constant system pressure.

Domestic water supplies are fed regularly to the system from the top 1 to 15 feet of water in elevated tanks. As the water level in the tank drops, the tank controls call for additional high-service pumps to start in order to satisfy the system demand and refill the tanks. The high-service pumps are constant-speed units, which can be operated at their highest efficiency point virtually all of the time. The remaining water in the tanks, which is roughly 70 percent to 75 percent capacity, normally is held in reserve as dedicated fire storage. This reserve will feed the system automatically as the fire flow demand and the domestic use demand at a specific time exceed the capacity of the system's high-service pumps.

Pumping for distribution storage: The two types of distribution storage, ground and elevated, have in turn, two types of pumping systems. One is a direct pumping system, in which the instantaneous system demand is met by pumping, with no elevated storage provided. The second type is an indirect pumping system in which the pumping station lifts water to a reservoir or elevated storage tank, which floats on the system and provides system pressure by gravity. Basic details on each type of system are presented below:

● Direct pumping: This system relates to older water systems and, with the exception of small communities, is generally very rare today, but some do exist. Variable-speed pumping units that operate off direct pressure are also in use in some medium sized communities. Hydropneumatic tanks at that pumping station provide some storage. These tanks permit the pumping station pumps to start and stop, based on a variable system pressure preset by controls operating off the tank.

● Indirect pumping: In an indirect pumping system, the pumping station with one or more pumps is not associated with the demands of the major load center on the water system. It is operated from the water level difference in the service or elevated storage cent(s), enabling the prescribed water level in the tank to be maintained. The majority of water systems today have an elevated storage tank or reservoir (or both) on high ground, floating on the water distribution system. This arrangement permits the pumping station to operate at a uniform rate, with the storage either making up or absorbing the difference between station discharge and system demand.

Analysis of storage: Two variations of distribution storage design affect the operation and reliability of a water system's capability to meet consumer consumption and fire suppression capabilities. These variations involve placement of the storage between the supply point and the major load center or beyond the major load center. An analysis of the following storage designs will be made in the remainder of this chapter:

● System A: a pumping station supplying the major center of the demand (load) with no elevated storage;

● System B: a pumping station supplying the major center with an elevated storage tank between the supply and the demand (load) point; and

● System C: a pumping station supplying the major center of demand with an elevated storage tank beyond the demand (load) point.

System Criteria

● Normal minimum working pressure in the distribution system should be approximately 50 psi and not less than 35 psi during a maximum hour. A normal working pressure in most systems will vary between 50 and 56 psi.

● System must be designed to maintain a minimum pressure of 20 psi at ground level at all fire hydrants on the distribution system under required fire flow conditions.

● Maximum day demand is 1.5 times the average daily demand.

● Maximum hour demand is 2.25 times the average daily demand.

Components of the Model System

The model system used in the analysis of the three systems (i.e., A, B, and C) has the following characteristics:

- ● population = 27,000;

- ● water demand rules:
    - Average Daily Consumption (ADC)   = 27,000 × 150 gpcd = 4.0 mgd,
    - Maximum Daily Consumption (MDC) = 4.0 × 1.5   = 6.0 mgd, and
    - Maximum Hour Consumption (MHC) = 6.0 × 1.5   = 9.0 mgd;

- ● needed fire flow (NFF):
    - Maximum NFF is 5,000 gpm   = 7.2 mgd, and
    - Maximum Day Rate plus NFF: 6.0 + 7.2 = 13.2 mgd; and

- ● pressure profile:
    - Minimum pressure at the major load center   = 50 psi, and
    - Minimum pressure on the maximum day plus fire flow = 20 psi

Special Note: System pipelines are all expressed as equivalent lengths of 24-inch pipe with a C factor of 120. The hydraulic gradient is the slope of the line joining the elevations to which water would rise in pipes freely vented and under atmospheric pressure.

System A: No storage: If no storage is provided in System A, as depicted in **Figure 4-13**, at a given demand rate, the pumping station hydraulic gradient must be sufficient to overcome system losses at a demand rate and maintain a minimum of 115 feet at the major load center. Thus, the pumping heads required maintaining 115 feet plus the head loss in 40,000 feet of equivalent pipe for the various conditions as follows:

Demand Rates		Pumping Head Required
Average day, 4.0 mgd:	115 + (0.67 × 40)	= 142 ft.
Maximum day, 6.0 mgd:	115 + (1.42 × 40)	= 172 ft.
Maximum hour, 9.0 mgd:	80 + (3.0 × 40)	= 200 ft.
Maximum day + NFF, 13.2 mgd:	46 + (6.1 × 40)	= 290 ft.

System B: Storage ahead of the load center: If, as shown in **Figure 4-12**, a 1.0-million-gallon storage tank is located 130 feet above datum plane and at a distance of 35,000 feet from the pump station (5,000 feet ahead of the major load center), the pumping head of a given pumping rate must be sufficient to pump against a head at the storage tank and overcome system losses at the pumping rate.

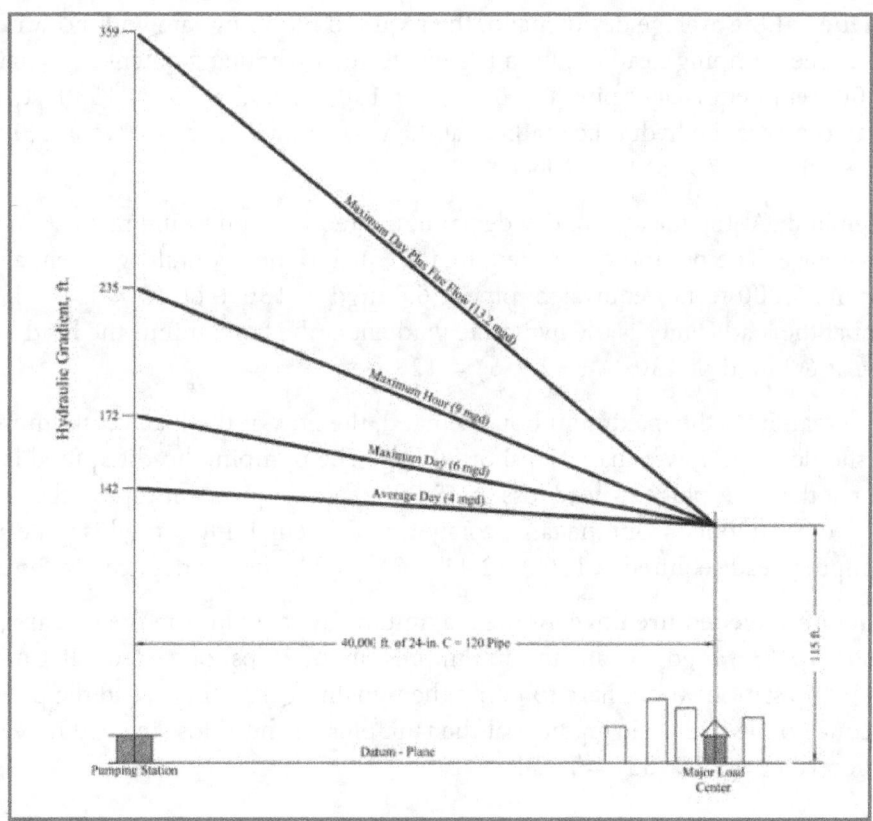

Figure 4-13: System A—Hydraulic Gradient with no Storage

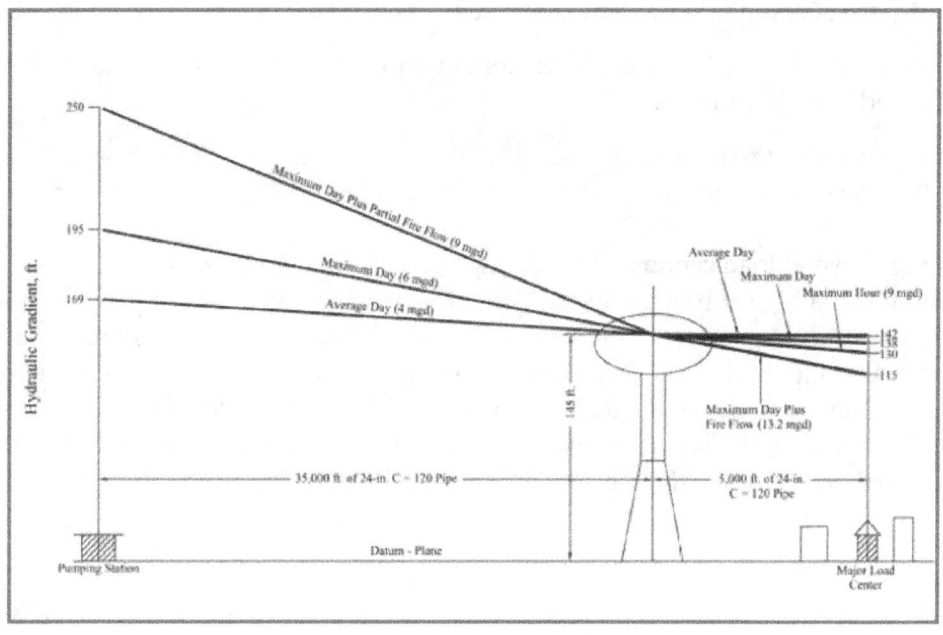

Figure 4-14: System B—Hydraulic Gradients with Storage Between Pump Station and the Load Center

Average day demand: At the average day demand, the required pumping rate with no water taken from storage is 4.0 mgd. The pumping head required is equal to the hydraulic gradient at the tank, plus the head loss in 35,000 feet of equivalent pipe at 4.0 mgd, or 130 + (0.67 × 35) = 153 feet. The hydraulic gradient at the load center is the hydraulic gradient at the tank, minus the head loss in 5,000 feet of equivalent pipe or 130—(0.67 × 5) = 127 feet.

Maximum day demand: At the maximum day demand, the required pumping rate is 6 mgd with no water taken from storage. The pumping head required is equal to the hydraulic gradient at the tank, plus the head loss in 35,000 feet of equivalent pipe at 6.0 mgd or 130 + (1.42 × 35) = 195 feet. The hydraulic gradient at the load center is the hydraulic gradient at the tank, minus the head loss in 5,000 feet of equivalent pipe at 6.0 mgd or 130—(1.42 × 5) = 123 feet.

Maximum hour demand: At the maximum hour demand, the flow at the load center must be 9 mgd and the pressure should be no lower than 35 psi or 80 feet. The pumping head required is equal to the hydraulic gradient at the tank, plus the head loss in 35,000 feet of equivalent pipe at the chosen pumping rate. If 1.0 mgd is to be supplied from the tank storage and the remaining 8 mgd is to be supplied from pumping, the pumping head required is 130 + (2.41 × 35) = 214 feet as depicted in **Figure 4-14**.

Maximum day rate plus needed fire flow: At the maximum day rate plus the fire demand, the flow to load center must be at a 13.3 mgd rate at a minimum pressure of 20 psi or 46 feet. If 1 mgd is supplied from storage, the pump station would have to pump the remaining 12.2 mgd, and the pumping head required is then equal to the hydraulic gradient at the tank plus the head loss in 35,000 feet of equivalent pipe at 12.2 mgd or 100 + (5.2 × 35) = 282 feet.

Demand Rates	Theoretical Pumping Head Required
Average day rate, 4.0 mgd: no water from storage	= 153 ft.
Maximum day rate, 6.0 mgd: no water from storage	= 180 ft.
Maximum hour rate, 9.0 mgd: 8.0 mgd rate from pump(s) and 1.0 mgd rate from storage	= 214 ft.
Maximum day rate plus fire flow, 13.2 mgd: 12.2 mgd rate from the pumps plus 1.0 mgd rate from storage	= 282 ft.

System C: Storage beyond load center: In the arrangement shown in **Figure 4-15**, 1.0 million gallons of storage is provided 5,000 feet beyond the load center, or 45,000 feet from the pump station, at an elevation of 119 feet above the datum plane. When no waste is being taken from storage at a given demand rate, the pumping head must be sufficient to pump against the head at the tank and overcome losses between the pump station and the load center at that demand rate. When part of the demand is being supplied from storage, the pumping head needs only to be sufficient to pump against the head at the load center and overcome losses in the pipeline between the pump station and the load center.

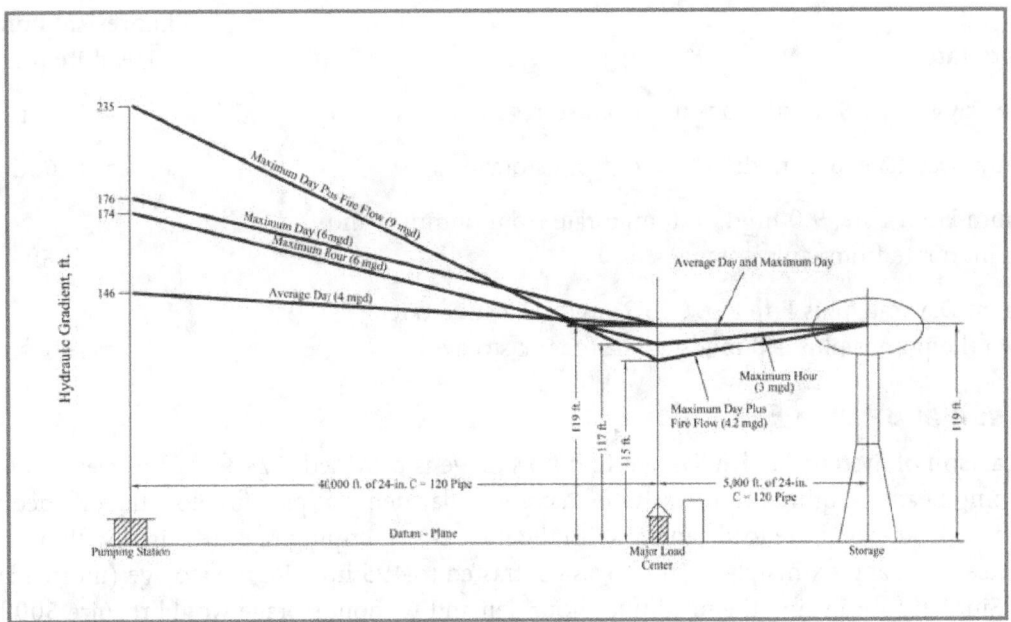

Figure 4-15: System C—Hydraulic Gradients with Storage Beyond the Load Center

Average day demand: At the average day demand, the required pumping rate is 4.0 mgd with no water taken from storage. The pumping head required is equal to the hydraulic gradient at the tank, plus the head loss in 40,000 feet of equivalent pipe or 119 + (0.67 × 40) = 146 feet. The hydraulic gradient at the load center is thus identified as that at the tank, or 119 feet.

Maximum day demand: At the maximum day demand, the required pumping rate is 6.0 mgd with no water taken from storage. The pumping head required is equal to the hydraulic gradient at the tank, plus the head loss in 40,000 feet of equivalent pipe at 6.0 mgd, or 119 + (1.42 × 40) = 176 feet. The hydraulic gradient at the load center is thus identified as that at the tank, or 119 feet.

Maximum hour demand: For a maximum hour demand which in the example is 9.0 mgd, the working pressure can be lowered to 35 psi or 80 feet of head. The storage tank could supply water at a rate of 700 gpm or 1.0 mgd during the hour and the remaining 8.0 mgd would be supplied from pumping. They hydraulic gradient at the load center is the hydraulic gradient at the tank, minus the head loss in the 5,000 feet of pipe between the tank and the load center at the storage discharge rate of 1.0 mgd, or 119—(0.05 × 5) = approximately 119 feet. The pumping head required is equal to the hydraulic gradient at the load center plus the head loss in 40,000 feet of equivalent pipe at 8.0 mgd, which is 119 + (2.41 × 40) = 215 feet.

Maximum day rate plus needed fire flow: The pressure at the load center can be lowered to 20 psi or 46 feet during a fire. Assuming the storage tank is full (1.0 mgd) this flow can move from the tank to the load center at 2,800 gpm or a 4.0 mgd rate. Thus, the remainder of the demand, which is 9.0 mgd, must be supplied by pumping. The pumping head required is equal to the hydraulic gradient at the load center or 46 feet plus the head loss in 40,000 feet of equivalent pipe, or 46 + (3 × 40) = 166 feet of head.

Demand Rates	Theoretical Pumping Head Required
Average Day Rate, 4.0 mgd: no water from storage	= 146 ft.
Maximum Day Rate, 6.0 mgd: no water from storage	= 176 ft.
Maximum Hour Rate, 9.0 mgd: 8.0 mgd rate from pump(s) and 1.0 mgd rate from tank storage	= 215ft.
Maximum Day Rate plus Fire Flow, 13.2 mgd: 9.2 mgd rate from the pumps plus 4.0 mgd rate from tank storage	= 166 ft.

Comparison of System Examples

The Comparison of System A with System C: If no storage is provided, 124 feet (290 feet—166 feet) more pumping head is required to furnish the maximum day demand plus fire flow than if adequate storage is provided beyond the load center. With the increased pumping rates required with no storage, the power needed is approximately 1,100 hp, as contrasted to 495 hp with the storage (more than twice as much). Similarly, furnishing the maximum hour demand without storage would require 500 hp, as opposed to 245 hp, still more than twice as much.

The capacities of the pumps required under these two conditions would be 13.2 mgd at 359 feet of head, as contrasted to 9.0 mgd at 235 feet of head, as opposed to 6.0 mgd at 174 feet of head. During average and maximum day demands, the pumping head at the source is approximately the same.

Comparison of System B with System C: In comparing storage located between the source and the load center with storage located beyond the load center, the example illustrates that pumping heads at the pumps are substantially lower under all conditions if the load center is located in the pumping station and the elevated storage.

Recommended Practice on Water Storage

The AWWA's *Manual of Water Supply Practices* (M-31) states that the recommended design is to have the major load center located between the pumping station and an elevated storage tank of sufficient capacity. This system is the most cost-effective design from a capital cost and operating cost standpoint. All those individuals that are involved in the planning, designing, and evaluating community water systems need to keep this concept in mind. It is not a rule but it is a good practical guide.

References:

1. American Water Works Association. *Manual on Water Supply Practices*, AWWA M-31, 1992 Ed. American Water Works Association, 6666 West Quincy Ave., Denver, CO 80235.

2. _____. *Distribution Operator Training Handbook*, 1996 Ed. American Water Works Association, 6666 West Quincy Ave., Denver, CO 80235.

3. Cote, A.E., and J.L. Linville, eds. *Fire Protection Handbook*, 16th ed. Quincy: National Fire Protection Association, 1986.

4. Fair, F.M., et al. *Water and Wastewater Engineering*. New York: John Wiley and Sons, Inc., 1966.

5. Steel, E.W., and T.G. McGhee. *Water Supply and Sewage*. New York: McGraw-Hill Book Co., 1989.

6. *Recommended Standards for Water Works, Great Lakes*. Upper Mississippi River Board of State Sanitary Engineers. Health Education Services. New York: New York State Health Department, Albany, 1987.

7. Ballou, A.F. "Hydrant Connections and Control of Dual Mains." *Journal NREWWA*, 52:81 (1938).

8. *Water Supply for Public Fire Protection, A Guide to Recommended Practice*. Toronto, Ontario: Fire Underwriters Survey, 1981.

9. Insurance Services Office. *Fire Suppression Rating Schedule*, 2003 Ed. Office of Natural Hazards Mitigation, ISO Building, 545 Washington Blvd., Jersey City, NJ 07310-1686.

10. National Fire Protection Association. *Fire Protection Handbook*, 18th ed. Quincy: author, 1997, pg. 6-40.

11. International Fire Service Training Association. *Water Suppliers for Fire Protection*, 3rd ed. IFSTA 205, p. 45.

CHAPTER 5: Consumer Consumption and Needed Fire Flow

The previous chapters have stressed the fact that community water systems need to provide an adequate and reliable source of treated water for consumer consumption and commercial use. In addition, water has to be made available for structural fire suppression and for other emergencies, such as hazardous material incidents, transportation vehicle fires, natural cover fuel fires, within a community that expose a community and water supplies for installed fire suppression systems, including automatic sprinkler systems. Because of economic considerations associated with the water system infrastructure, all but the very largest cities have water systems that provide one water system for consumer consumption and fire protection. As reflected in Chapter 4, water demand needs to be provided during all periods of each day, day of the week, and week of the year for consumer use plus the capability to meet required fire flows when a serious fire erupts in a community. The question now to be answered is how much water is needed for domestic consumption and how much is required to meet needed fire flows at representative locations throughout a community such as commercial, industrial, and residential areas. This chapter provides guidelines estimating water supply needs.

The basic question to be answered by a community water planning board or commission is "how much water will our system be required to deliver and to where, both today and in the future?" The answer to this question will require the acquisition of basic information about the community, including historical water usage, population trends, planned growth, topography, and existing system capabilities, to name just a few. This information can then be used to plan for the logical extension of the existing system and to determine needed improvement to provide sufficient water supply at demand pressures throughout the community.

Water Demands

The first consideration of a water distribution system is the determination of the quantity of water that will be required, with provision for the estimated requirements for the future.

In terms of total quantity for domestic consumption, the water demand in a community usually is estimated on the basis of per capita demand. According to a study published by the U.S. Geological Survey, the average was estimated to be about 105 gallons per day per capita in 1990. However, by the year 2000, the AWWA estimated the per capita use was 140 gallons per day. Water demand has been demonstrated to vary rather significantly by State as reflected in the referenced 1990 study. This is illustrated in **Table 5-1**.

Table 5-1: Estimated Use of Water in the United States in 1990

State	Gallons per Capita per Day
Alabama	150
Alaska	106
Arizona	147
Arkansas	145
California	70
Colorado	78
Connecticut	179
Delaware	111
District of Columbia	115
Florida	119
Georgia	186
Hawaii	90
Idaho	76
Illinois	66
Indiana	86
Iowa	70
Kansas	124
Kentucky	58
Louisiana	105
Maine	66
Maryland	77
Massachusetts	148
Michigan	123
Minnesota	86
Mississippi	129
Missouri	115
Montana	213
Nebraska	71
Nevada	75
New Hampshire	135
New Jersey	119*
New Mexico	67
New York	86
North Carolina	50
North Dakota	85
Ohio	100
Oklahoma	79
Oregon	111
Pennsylvania	62

continued on next page

State	Gallons per Capita per Day
Rhode Island	67
South Carolina	76
South Dakota	81
Tennessee	85
Texas	143
Utah	218
Vermont	80
Virginia	75
Washington	138
West Virginia	74
Wisconsin	52
Wyoming	163
Puerto Rico	48
Virgin Islands	23
United States Total Average	**105**

*Excluding New York City

Ref.: Mays, Larry W. Water Distribution Systems Handbook. New York: McGraw-Hill, 1999.

The water usage shown above illustrates a wide variation. Per capita use varies from a low use in Maine of 58 gallons per day capita to Utah with 218 gallons per day per capita. These variations depend on geographical location, climate, size of the community, extent of industrialization, and other influencing factors unique to most communities. Because of these variations, the only reliable way to estimate future water demand is to study each community separately, determining existing water use characteristics and extrapolating future water demand using population trends.

In terms of how the total water use is distributed within a community throughout the day, perhaps the best indicator is land use. In a metered community, the best way to determine water demand by land use is to examine actual water usage for the various types of land uses. The goal of examining actual water usage is to develop water "duties" for the various types of land uses that can be used for future planning. Water duties are normally developed for the following land uses:

- Single-family residential—Some communities have low, medium, and high density zones.
- Multifamily residential.
- Commercial—Normally divided according to office and retail categories with a separate category for shopping plazas.
- Industrial—Normally divided into light and heavy categories and separate categories for high-end users.
- Public buildings—Normally include courthouses, libraries, post office, police stations and jails, community and senior centers, fire stations, fire training areas, and elementary or high schools.
- Public property—Normally includes parks, grassed areas that are watered, swimming pools, etc.
- Special properties—Normally includes Federal and State buildings, colleges and universities, local airports, and any other properties not identified above.

Water duties normally are expressed either by specific establishments or in gallons per acre per day. It may be wise to conduct parallel studies to examine any differentials for the community at large. Differences can be examined by specific land usage.

Table 5-2 presents water use for many different establishments. Although the rates may vary widely, they are useful in estimating total water use for individual users when no other data are available.

Table 5-2: Typical Rates of Water Use for Various Establishments

User	Range of Flow Gallons per Person on Units per Day
Airport, per passenger	3–5
Assembly hall	2–3
Bowling alley, per alley	16–26
Camp	
Pioneer type	21–32
Children's w/central toilet and bath	42–53
Day, no meals	42–53
Luxury w/private bath	79–106
Labor	37–53
Trailer w/private toilet and bath per night	132–159
Country clubs	
Residential type	79–159
Transient type serving meals	16–26
Dwelling unit, residential	
Apartment house on individual well	79–106
Apartment House on public water supply	79–132
Note: higher if unmetered	
Boarding house	40–58
Hotel	53–106
Lodging house and tourist home	32–53
Motel	106–159
Private dwelling on an individual well	53–159
Note: use same values for metered supply	
Private dwelling on a nonmetered public water supply	106–211
Factory, sanitary wastes, per shift	40–100
Fairground (based on daily attendance)	1–2
Institution	
Average type	106–159
Hospital	185–317
Office space	11–16
Park and recreational facilities w/flush toilets	5–11

continued on next page

User	Range of Flow Gallons per Person on Units per Day
Restaurants w/toilet facilities	
Average	7–11
Kitchen waste only	3–5
Short order	3–5
Bar and cocktail lounge	2–3
24 hour seating facilities, per seat	42–58
Tavern per seat	16–26
Schools	
Day w/cafeteria and lunchroom	11–16
Day w/cafeteria and showers	16–21
Boarding	53–106
Self-serviced laundry, per machine	264–793
Store	
Mercantile w/25 feet of frontage	423–528
Each additional 25 feet of frontage	370–423
Swimming pool and bench w/toilet and shower	11–16
Theater	
Indoor, per seat, 2 showings per day	3–5
Outdoor, per car, including food stand	3–5

Ref.: Mays, Larry W. *Water Distribution Systems Handbook*. New York: McGraw-Hill, 1999.

Table 5-3: Typical Water Duties

Land Use	Water Duty (gal/day) (gal/day/acre)		
	Low	High	Average
Low-density residential	400	3,300	1,670
Medium-density residential	900	3,800	2,610
High-density residential	2,300	12,000	4,160
Single-family residential	1,300	2,900	2,300
Multifamily residential	1,100	6,600	4,160
Office–Commercial	1,100	5,100	2,030
Retail–Commercial	1,100	5,100	2,040
Light–Industrial	200	4,700	1,620
Heavy–Industrial	200	3,100	2,270
Parks	400	3,100	2,020
Schools	400	2,500	1,700

Ref.: Adapted from Montgomery Watson Study of Date of 28 Western U.S. Cities.

A Suitable Approach for Studying Consumer Use Water Demand

A large-scale community map is needed. The actual scale has to be determined by the typical distance between street intersections in a community. Identify the class of properties along both sides of the street between intersection streets, or to dead ends or where a street leaves the corporation line. Using the guidelines from **Tables 5-2** and **5-3**, estimate the gallon per day rate along street fronts or use contiguous blocks to form a square or rectangle with approximately 43,000 square feet which is generally equivalent to an acre. Plot this information on the street map and sum the gallons of water required for the community for normal daily consumption.

Fire Suppression Water Demand

Criteria for establishing needed fire flows: an adequate amount of water is essential for confining, controlling, and extinguishing hostile fires in structures. The actual amount of water needed differs throughout a community due to different building and occupant conditions. Therefore, water demand for structural fire protection has to be determined at a number of different locations throughout a given community or fire protection district established at the local or county level of government in the United States.

The ISO evaluates the fire suppression capabilities of communities according to 10 classifications of service. These classifications are conducted as part of the fire and property insurance rating system for over 130 insurance companies writing this type of insurance in the United States. Three protection features are evaluated, and a percent of credit is assigned to each feature as follows:

1) Fire Department   50 percent
2) Water Supply   40 percent
3) Receiving and Handling Fire Alarms   10 percent
             Total   100 percent

The weight given to water supply stresses the need for an adequate water supply for structural fire suppression. The water supply evaluation considers the ability of the water system to meet consumer consumption plus fire flow under average and maximum daily conditions. Approaches to evaluating a municipal water system are presented in a companion document titled *Evaluating Municipal Water Supply System*. The following material is restricted to the determination of needed fire flows at representative locations throughout the community. These locations are selected by the ISO to represent both typical and special fire risks including commercial, mercantile, industrial, institutional, and residential properties for insurance rating purposes.

Needed fire flow is the amount of water considered necessary to control a major developing fire in a specific building. The amount of water needed is expressed in gpm at 20 psi residual pressure for a duration ranging from 2 to 4 hours. The minimum needed fire flow for any single building is 500 gpm for 2 hours. The maximum needed fire flow is 12,000 gpm for 4 hours.

Needed fire flow at selected sites throughout a given community is intended as one measure used to compare and contrast available fire flow to a specification fire flow. Studies conducted by insurance industry over many years shows a high statistical correlation between the amount of water actually used to confine, control, and extinguish a dingle building fire and the amount of water determined form the application of the needed fire flow analysis. It is important to recognize that a fire that spreads beyond the building or origin requires larger amounts of water for longer periods of time. The ISO Fire Suppression Rating Schedule (FSRS or Grading Schedule) provides one analytical method for determining capability to suppress a developing fire in representative buildings throughout a given community.

Water supply studies further indicate it is unusual for an existing water distribution system to provide sufficient volume and pressure to meet every SFF. However, in designing a new water supply and distribution system or expansion to an existing system, or improvements to components of an existing water supply, pipe network and hydrant installation, plans should consider water delivery rates and pressure profiles to provide needed fire flow at specific locations throughout a community as determined by the ISO. Upon request, the ISO can provide a listing of needed fire flows at representative locations for a given community. (1) In meeting fire flow objectives, a water supply system should be capable of delivering the maximum needed fire flow within the built-upon area served by the water system.

Basic Needed Fire Flow Formula: A needed fire flow for an individual building is computed from the formula below: (2)

$$NFF_i = (C_i) \times (O_i) \times [1 + P_i]$$

The calculation of needed fire flow, in gallons per minute (gpm), for a subject building, considers the construction (C_i), occupancy (O_i), exposure (X_i), and communication (P_i) factors of that building or fire division as outlined below.

- **Construction factor (C_i):** This portion of the needed fire flow attributed to the type of construction and square footage of the subject building is determined by the following formula from the ISO Grading Schedule.

  $C_i = 18F (A_i) (0.5)$ where:

  F = coefficient related to the class of construction as follows:

  - F = 1.5 for construction Class 1 (frame)

  - F = 1.0 for construction Class 2 (brick-joist or joisted masonry)

  - F = 0.8 for construction Class 3 (noncombustible) and construction Class 4 (masonry—noncombustible)

  - F = 0.6 for construction Class 5 (modified fire resistive) and construction Class 6 (full fire resistive)

  A_i = effective area (the total square foot of the largest floor in the building plus the following percentage of other floors:

  - For buildings of construction Class 1 to 4: 50 percent of all other floors

  - For buildings of construction Class 5 and 6: if all vertical openings in the building have 1 to 1-1/2 hour or more protection, 25 percent of the area, not exceeding the two largest floors. The doors shall be automatic or self-closing and labeled as class B fire doors. In other buildings, 50 percent of the area, not exceeding eight floors.

[Sidebar: ISO has additional explanatory information on calculating the effective area. There is no need to know this information, since ISO will provide such details on any specific building upon request. The intent of discussing this formula is to make the reader aware of the factors that influence the needed fire flow calculation process.]

The maximum value of C_i is limited to the following:

- 8,000 gpm for construction Class 1 and 2
- 6,000 gpm for construction Class 3 to 6
- 6,000 gpm for a one-story building of any class of construction

The minimum value of C_i is 500 gpm. The calculated value of C_i should be rounded to the nearest 250 gpm.

- ● **Occupancy factor (O_i):** given in **Table 5-4**, reflects the influence of the occupancy in the subject building on the needed fire flow. Representative lists of occupancies by combustibility class are given in **Figures 5-1** and **5-2**.

- ● **Exposures (X_i) and communications (P_i) factors:** The exposures and communications factors reflect the influence of exposed and communicating building on the needed fire flow. A value for ($X_i + P_i$) shall be developed for each side of the subject building as follows:

$$(X + P)_i = 1.0 + \sum_{i=1}^{n} (X_i + P_i), \text{ maximum } 1.75$$

where: n = number of sides of subject building

The factor for X_i (exposure) depends on the construction and length-height value (length of wall in feet times height in stories) of the exposed building and the distance between facing walls of the subject building and the exposed building. This factor shall be selected from **Table 5-5**.

The factor for P_i (communications) depends on the protection for communicating party wall openings and the length and construction of communications between fire divisions. The factor shall be selected from **Table 5-6**. When more then one communications type exists in any one side wall, apply only the largest P_i for that side. When there is no communication on a side, $P_i = 0$.

continued on next page

Classification 1

Steel or concrete products storage, unpackaged

Classification 2

Apartments	Hotels
Churches	Motels
Court houses	Offices
Dormitories	Parking garages
Hospitals	Schools

Classification 3

Amusement park buildings, including arcades and game rooms
Automobile sales and services
Department stores
Discount stores
Food and beverage—sales, service, or storage
General merchandise—sales or storage
Hardware, including electrical fixtures and supplies
Motion picture theaters
Pharmaceutical retail—sales and storage
Repair or service shops
Supermarkets
Unoccupied buildings

Classification 4

Aircraft hangers—with or without servicing/repair
Auditoriums
Building material—sales and storage
Freight depots, terminals
Furniture—new or secondhand
Paper and paper product—sales and storage
Printing shops and allied industries
Theaters—other than motion picture
Warehouses
Wood product—sales and storage

Classification 5

Chemical sales and storage
Cleaning and dying material—sales and storage
Paint—sales and storage
Plastic or plastic product—sales and storage
Rag sales and storage
Upholstering shops
Waste and reclaimed material—sales and storage

Figure 5-1: Typical Occupancy Classifications— Nonmanufacturing

Classification 2

Ceramics manufacturing
Concrete or cinder products manufacturing
Fabrication of metal products
Primary metals industries

Classification 3

Baking and confectionery
Dairy processing
Leather processing
Soft drink bottling
Tobacco processing

Classification 4

Apparel manufacturing
Breweries
Cotton gins
Food processing
Metal coating and finishing
Paper product manufacturing
Rubber product manufacturing
Woodworking industries

Classification 5

Cereal and flour mills
Chemical manufacturing
Distilleries
Fabrication of textile products (except wearing apparel)
Meat or poultry processing
Plastic products manufacturing
Textile manufacturing

Figure 5-2: Typical Occupancy Classifications—Manufacturing

continued on next page

Table 5-5: Factor for Exposure (Xi)

Construction of Facing Wall of Subject Bldg.	Distance to Exposed Bldg.	Length-Height* of Facing Wall of Exposed Bldg.	Construction of Facing Wall of Exposed Building			
			Construction Classes			
			1,3	2,4,5,6	2,4,5,6	2,4,5,6
				Unprotected Openings	Semiprotected Openings (wire glass or outside open Sprinklers)	Blank Wall
			Exposure Factor Xi			
Frame, metal, or masonry, w/ openings	0–10 ft	1-100	0.22	0.21	0.16	0
		101–200	0.23	0.22	0.17	0
		201–300	0.24	0.23	0.18	0
		301–400	0.25	0.24	0.19	0
		Over 400	0.25	0.25	0.20	0
	11–30 ft	1–100	0.17	0.15	0.11	0
		101–200	0.18	0.16	0.12	0
		201–300	0.19	0.18	0.14	0
		301–400	0.20	0.19	0.15	0
		Over 400	0.20	0.19	0.15	0
	31–60 ft	1–100	0.12	0.10	0.07	0
		101–200	0.13	0.11	0.08	0
		201–300	0.14	0.13	0.10	0
		301–400	0.15	0.14	0.11	0
		Over 400	0.15	0.15	0.12	0
	61–100 ft	1–100	0.08	0.06	0.04	0
		101–200	0.08	0.07	0.05	0
		201–300	0.09	0.08	0.06	0
		301–400	0.10	0.09	0.07	0
		Over 400	0.10	0.10	0.08	0
Blank masonry wall	When facing wall of the exposed building is higher than subject building, use the above information, except use only the length-height of facing wall of the exposed building above the height of the facing wall of the subject building. Buildings five stories or over in height, consider as five stories. When the height of the facing wall of the exposed building is the same or lower than the height of the facing wall of the subject building, $Xi = 0$.					

continued on next page

Table 5-6: Factor for Communications (Pi)*

Protection of Passageway Openings	Fire-Resistive, Non-Combustible or Slow-Burning Communications				Communications with Combustible Construction					
	Open	Enclosed			Open			Enclosed		
	Any Length	10 ft. or less	11 to 20 ft.	21 to 50 ft.	10 ft. or less	11 to 20 ft.	21 to 50 ft.	10 ft. or less	11 to 20 ft.	21 to 50 ft.
Unprotected	0		0.30	0.20	0.30	0.20	0.10			0.30
Single Class A (fire door at one end of passageway)	0	0.20	0.10	0	0.20	0.15	0	0.30	0.20	0.10
Single Class B (fire door at one end of passageway)	0	0.30	0.20	0.10	0.25	0.20	0.10	0.35	0.25	0.15
Single Class A (fire door at each end or double class A fire doors at one end of passageway)	0	0	0	0	0	0	0	0	0	0
Single Class B (fire door at each end or double class B fire doors at one end of passageway)	0	0.10	0.05	0	0	0	0	0.15	0.10	0

Source: *Copyright, ISO Commercial Risk Services, Inc.*, 1980.

Notes: 1.   The reader should refer to the 1980 Fire Suppression Rating Schedule, published by ISO Commercial Risk Services, New York (1980), for complete information regarding factors for communication.

      2.   When a party wall has communicating openings protected by a single automatic or self-closing class B fire door, it qualifies as a division wall for reduction of area.

      3.   Where communications are protected by a recognized water curtain, the value of Pi is 0.

* The factor for Pi depends on the protection for communicating party wall openings and the length and construction of communications between fire divisions. Pi shall be selected from this table. When more than one communication type exists in any one side wall, apply only the largest factor Pi for that side. When there is no communication on a side, Pi = 0. (Party wall means a division wall rated 1 hour or more with labeled class B fire doors on openings.)

For over 50 ft, Pi = 0

For unprotected passageways of this length, consider the two buildings as a single fire division.

Special Notes on the Determination of Needed fire flows

1) When a wood-shingle roof covering on a building or on exposed buildings can contribute to spreading fires, add 500 gpm to the needed fire flow.

2) The needed fire flow shall not exceed 12,000 gpm nor be less than 500 gpm.

3) The needed fire flow shall be rounded to the nearest 250 gpm if less than 2,500 gpm, and to the nearest 500 gpm if greater than 2,500 gpm.

4) For one- and- two family dwellings not exceeding two stories in height, the needed fire flow listed in **Table 5-7** shall be used.

Table 5-7: Needed fire flow for One-and Two- Family Dwellings*

Distance Between Buildings Feet	Needed fire flow gpm
Over 100	500
31–100	750
11–30	1,000
Less than 11	1,500

*Dwellings not to exceed two stories in height.

[Side bar: Many communities today are allowing contractors in subdivisions to build one family homes 20 feet apart. Note that this would require a needed fire flow of 1,000 gpm for all residential areas with this separation distance because of the identified exposure problem.]

5) The water supply for buildings protected by automatic sprinkler systems that are in compliance with the NFPA Standard 13 and inspected semiannually are "coded" as sprinkler buildings for fire insurance rating purposes. The water demand requirement for these buildings is determined from the stated NFPA standard including the allowance for hose steams where required.

References:

1. American Water Works Association. *Distribution System Requirements for Fire Protection*, M-31. American Water Works Association, 6666 West Quincy Ave., Denver, CO 80235.

2. Insurance Services Office. *Fire Suppression Rating Schedule*, 2003 Ed. Insurance Services Office, Office of Natural Hazards Mitigation, ISO Building, 545 Washington Blvd., Jersey City, NJ 07310-1686.

CHAPTER 6: Alternative Water Supplies

People living in local communities generally assume that there is a water supply system that will meet their domestic needs, and that there will be water available from fire hydrants to fight building fires. (1) Concurrently, people living outside incorporated areas and in rural areas do not generally expect to have water mains that supply residential and commercial property. Water supply for structural firefighting was limited to that carried on the first alarm response to a given fire site. Water supply for the urban landscape and the rural landscape was pretty well divided until the late 1970's. Today's suburban and rural growth patterns are making this dichotomy on water supply very fuzzy.

The grow pattern of homes, schools, and commercial property outside of incorporated villages, towns, and cities requires a new understanding of water supply for consumer consumption, the adequacy and reliability of developed water supplies for both consumer consumption and fire protection. These newly built-upon land areas may have one of the following arrangements for meeting water supply needs:

- Each occupied building is supplied by well water on the owner's land site; there are no water mains or public fire hydrants. This method of water supply applies to single family dwellings, schools, small commercial property such as a convenience store, etc.

- Both commercial and housing developers may take a tract of land and sell building lots. The developer also may be the builder, or there may be several contractors available for building on the individual lots. Generally there is a small water system to meet the domestic consumption needs. The underground piping may be 2-inch to 3-inch pipe along street fronts, and taps are made on these lines to feed each home or small business. The water supply is typically from wells that is treated and pumped into to a ground-level or elevated storage tank to flow by gravity on demand through the pipe lines. The size of the water mains is inadequate to supply water for fire protection.

- The above description of land development may be modified in the following way. The developer plans that "someday" the growing community will be large enough to require a higher yield water source, and the associated need for a water treatment plant. These kinds of development often are planned to be located near a river, fresh water lake, or a good flow stream where a low-level dam can be erected to form a holding basin or reservoir of water for the water system. Recognizing this potential need, a decision is made to install fire hydrants in the development to save installation costs in the future. These hydrants then will be connected to 6-inch or larger pipe that will be laid in the same trench as the domestic piping system. The futuristic drawings and plans look good on paper. However, the installation of the fire hydrants may not follow the plan. There are several document cases where the installed fire hydrants are fed by the domestic system, the small-diameter pipe. When questioned about this practice, the response is often that the fire hydrants are used to water the green areas, trees, and to keeps the streets clean. However, this gives a false sense of security for the property owners where this practice is in place. The water supply simply is inadequate for fire protection, and it may be inadequate for these adjunct services in the community.

All of these noted water supply arrangements are of much interest and concern to property and casualty insurance companies that write insurance policies with coverage for fire loss. The ISO represents over 160 companies in the United States that write homeowners and commercial property fire insurance policies and packaged insurance policies that cover fire loss. In an ISO 2005 publication on *Effective Fire Protection—A National Concern*, there is startling information on the lack of community water supplies in fire districts across the country. For insurance rating purposes, the ISO has identified 48,515 fire districts or insurance-graded areas in the United States. Each area is evaluated by the ISO for fire suppression effectiveness using criteria developed for three needed fire protection services: 1) Receiving and Handling Fire Alarms, 2) Fire Department, and 3) Water Supply. Water supply accounts for 40 percent of the total evaluation. A *Grading Evaluation* conducted by ISO Field Representatives results in establishing a community or fire district PPC on a Scale of 1 to 10. Out of all the fire districts in the Nation, there are only 52 cities, communities for fire districts (i.e., a county might be a fire district based on local government administration and management).

From a water supply perspective, it is important to identify and consider the Class 9 and Class 10 properties. Class 10 property, for insurance rating purposes, is located beyond 5 travel miles for the first-due engine company responding from an ISO-recognized fire department, and there is **no** recognized water supply. The message is that there can be no effective fire protection if there is not a sufficient fire apparatus response time and personnel to fight fires, and there is insufficient water to extinguish fires. The likelihood of a total property loss from a developing fire in a structure in Class 10 *Graded Area* is statistically high. Generally, most fire insurance carriers will not underwrite fire insurance policies where there is an ISO Classification of 10. Some insurers will write this property at a very high premium. The latest ISO studies indicates that 2.9 percent of the Nation's fire districts receive a Class 10 designation.

In contrast, a Class 9 is assigned to a *Graded Area* structural property is within 5 travel miles of the first-due fire department engine company, but there is **no** recognized water supply. The same study referred to above indicates that 32.2 percent of the Nation's fire districts receive a Class 9 designation. The bottom line is that in this classification there is reasonable response from a fire company, but the water supply to fight a fire does not meet recognized criteria for the very large majority of the insurance industry. There is a defined probability of loss reduction with Class 9 property over Class 10 property. This property group includes farmstead property; isolated homes away from a community; and vacation homes on lakes, in the mountains, or at the sea-shore, that are not near a municipal water system with fire hydrants.

It is of a major concern to realize that, by combining the Class 9 and Class 10 fire districts together, 35.1 percent of the entire Nation's fire districts do not have a recognized water supply!

Now, it is essential to define a recognized water supply by the insurance industry. The ISO *Fire Suppression Rating Schedule*, or *Grading Schedule* for short, states that the minimum recognized water supply for structural fire protection is 250 gpm for a 2-hour duration; the minimum volume of water is 30,000 gallons. However, the minimum water supply for any single structure is 500 gpm for 2 hours. Prorated credit is given if only 250 gpm is available; no credit is given if a structure is not protected by 250 gpm for 2 hours.

All of the above means that 17,029 fire districts in the United States either have no municipal water supply system, or the existing water supply system does not have the capability to supply 250 gpm for 2 hours.

So, if this manual is restricted to discussing municipal water supply systems, it leaves out over one-third of the Nation's fire districts. Therefore, it is essential to discuss alternative water supplies to meet fire protection needs. Alternative water supply methods are essential to augmenting existing municipal water supplies where the existing system cannot meet needed fire flows. Furthermore, alternative water supply methods can be used where a water supply system's service has been interrupted for whatever reason. Some of those reasons include storm damage, power failures to pumping equipment, drought conditions where there just is no water available, and, after 9-11, the ever-present threat of terrorism or simple sabotage to a water system. Therefore, this chapter is devoted to presenting the major concepts of alternative water supplies as developed through a partnership project between the U.S. Fire Administration (USFA) and the ISO.

Alternative Water Supply Concepts

The primary objective of alternative water supplies is for fire district or ISO *Graded Areas* that do not have recognized water supplies to plan and establish a water delivery program that will deliver a minimum of 250 gpm for 2 hours to all structural property within the fire district. There are three adjunct objectives to alternative water supply programs:

1) Raise the bar and make the minimum water delivery capability to all structural property in a given fire district or *Graded Area* 500 gpm. This will significantly increase the ability of responding fire departments to confine, control, and extinguish developing fires in structures, and will improve the credit given by the ISO for water supply significantly when conducting a PPC evaluation of the *Graded Area*.

2) Use developed alternative water supply delivery programs to augment existing municipal water supplies at the site of representative fire risks where there is a deficiency in meeting needed fire flows. Many municipal water systems, especially in the small communities, cannot meet needed fire flows on dead-end water mains and near the perimeter of the water distribution system. There also may be a deficiency in the capability of a given water supply system to meet high fire flow demands in the central business district and at specific fire risks such as schools and hospitals in the community.

3) Alternative water supplies can be used for emergency water supply in case there is disruption to any part or all of a municipal water system in order to maintain fire protection capability. There is the growing potential for alternative water supplies to be used for domestic consumption capability in the case that a municipal water system is down for 1) the lack of sufficient intake water supply due to drought conditions in the area, 2) due to equipment failures for a variety of reasons including power outages, and 3) domestic water supply capability when used with portable water supply purification equipment. In this later case, the military has in place water purification equipment to change seawater into drinking water. Now, the technology is in place that uses reverse osmosis processes and new technology for membrane filtering so that mobile filtration units can take any brackish water or salt water and treat this water to EPA standards. Such a unit has been purchased by a county fire department on Long Island, New York. This technology may be important to homeland security when addressing possible terrorist threats, or to the Federal Emergency Management Agency (FEMA) when considering disasters of the magnitude of Hurricane Katrina.

Where Are Alternative Water Supplies Needed?

Alternative water supplies are needed for the fire protection structural property that is not within 1,000 feet of a recognized fire hydrant (i.e., a fire hydrant that has the demonstrated capability to delivery a minimum of 250 gpm for a 2-hour duration). A scaled fire district map is needed to identify structural fire risks that are beyond 1,000 feet of a recognized fire hydrant. This may require running fire-flow tests to determine the capability of fire hydrants on the perimeter of water systems especially in smaller communities.

Water Supply Officer

Fire districts or fire departments that are serious about establishing an alternative water delivery program need to appoint a Water Supply Officer to coordinate *all* activities associated with the delivery of water supplies to fire risks throughout the fire area, to augment any existing water supplies that have deficiencies in meeting needed fire flows, and to meet special emergency needs using alternative water supplies. It should be understood by the appointing fire protection organization and the area local government that this individual has the authority and responsibility to organize, implement, and maintain the alternative water supply program. Often this assignment is given to an assistant fire chief as part of the training for effective incident command.

The individual selected for the water supply officer position must be trained to have knowledge about all aspects of the alternative water supply program, and be able to communicate effectively about program elements to fire department personnel, local and State authorities, and the public at large. Typical task performance for alternative water supply officers is referenced in the USFA/ISO *User Workbook and Alternative Water Supplies* noted with the chapter References. Additional knowledge and proficiency comes from actually establishing a water delivery program and through consulting other appropriate references at the end of this chapter.

Identifying and Certifying Alternative Water Supplies

It needs to be reinforced that the ISO's *Fire Suppression Rating Schedule* documents a minimum recognized available fire flow of 250 gpm for a duration of 2 hours for structural property. This applies whether a structural fire risk is served by a municipal water supply system providing water for consumer use or not. In suburban and rural areas without water delivery from a municipal water system supplying fire hydrants, provisions need to be made by the jurisdictional fire department to transport water oat structural fires and other fire emergencies using fire pumpers, pumper-tankers, and mobile water tankers. To meet the minimum specified flow of 250 gpm for 2 hours or a higher flow as determined by the needed fire flow for specific fire risks, water supply sources need to be provided throughout a fire protection district or *Graded Area*.

These water supply sources are divided into two categories: 1) natural water sources and 2) developed water sources. Several subclassifications of water sources are identified under each category outlined below:

Natural Water Source Classifications	*Developed Water Source Classifications*
Ponds	Cisterns
Lakes	Rivers
Streams	Driven wells
Reservoirs	Aboveground storage tanks
Streams	Belowground storage tanks
Rivers	Elevated storage tanks
Irrigation channels	Swimming pools
	Domestic water systems

1. **Factors associated with the reliability of water sources.**

   • Natural water sources are susceptible to drought conditions in most parts of the country.

   • Ground-level water sources in rural areas may be used by farmstead property owners to feed animals and water crops.

   • Farm ponds, lakes, streams, and rivers may freeze to varying levels during periods of severe cold weather.

   • Silt and debris can accumulate in a pond, lake, or reservoir, reducing the actual volume of water available for fire while the surface level of the water source remains constant.

   • Streams and rivers may reach flood stage and overflow the retaining banks making the water source nonaccessible.

2. **Factors of accessibility.**

   • All natural water sources used for fire protection should be accessible by all-weather roads and adequately plowed during and after snowfall in snow-belt areas.

   • The planning of new dedicated ponds and small lakes for fire protection should consider placement near well-traveled roads to permit filling mobile tankers for water shuttle operations.

   • A dry hydrant location, a drafting site, or a water fill site needs a constructed pad where a mobile pumper can draft water.

   • Bridges used for water supply access require special considerations. Fire apparatus cannot park on bridges to fill mobile water tankers in several States by law because the apparatus would block traffic. In many cases, it is best to have a dry hydrant fill point away from a bridge because this will reduce the water lift height and it will remove a very heavy weight from the bridge in rural areas. Fire departments need to check with the specific State Department of Transportation and abide by he laws governing this situation.

   • Access to fire sites and water fill sites using routes with bridges, especially in rural areas, needs to be carefully evaluated. A very large number of bridges were built for farm-to-market use. Now they carry increased traffic loads with the expansion of rural and suburban growth. NFPA 1142, Standard on Water Supplies for Suburban and Rural Fire Protection states that, "the general condition of bridges in the rural areas of most states is poor." The nature and extent of the rural and suburban bridge problem makes it a serious consideration when planning alternative water supply programs. This is an especially important consideration when planning transportation routes for fully loaded mobile tankers.

   • Access to private water supplies needs to be evaluated. Fire departments should establish a water usage agreement with the owner(s) of water supplies located on private property to gain access to water sources before the occurrence of a fire. Written agreements should be made in close cooperation with the municipal town or county attorney, Furthermore, it is highly desirable that the agreement be reviewed by a representative of the highway or the county road department or other person who will build, service, and maintain the access area or road leading to the water supply; this includes persons who will perform such functions as snow plowing in specific areas of the country. The property owner should have a copy of the agreement that will be used by different fire departments with approval of their county, township or political jurisdiction.

   • Alternative water supply sources should be provided at location to service structural property. A basic guideline to follow is that mobile tanker apparatus should not have to travel more than 2 miles to reach a recognized water source for refilling the tanker. However, the number of mobile water tankers available to respond to a given fire site will affect this guideline.

- Identification of water resources is most important. The responsible water supply officer should install or cause to be installed an appropriate reflective sign at each water supply point identifying the site for fire department use and including a site number for each recognized water source. NFPA recommends that this number be 3 inches high with a 2-inch stroke.

- Water maps should be prepared with access locations. The responsible water supply office should maintain a map showing the location and amount of water available at each water site. A copy of this map should be located in the jurisdictional fire alarm office or Emergency Operations Center (EOC), and a copy should be carried in the first-due engine company in the Graded Area (i.e., pumper or pumper-tanker) that responds to areas beyond recognized fire hydrants. All area fire chiefs' vehicles and the water supply officer's emergency response vehicle also should carry a current map along with the necessary prefire plans to implement an alternative water supply delivery system throughout the fire protection jurisdiction or Graded Area of responsibility.

3. **Reliability of Natural Water Supplies.**

- Impounded water supplies: These supply sources consist of naturally developed ponds in low-lying areas that are accessible by all-weather roads, constructed ponds for fire protection, and small lakes that are natural or manmade with a retention dam at the outflow location. NFPA 1142, 2002 edition, states that the quantity of water considered available for natural water supplies "is the minimum available during a drought with an average 50 year frequency that has been certified by a Professional Engineer or Hydrologist, or other similar qualified person." (__)

- Flowing streams: For all flowing streams, (i.e., rivers, streams, and creeks) NFPA 1142 states, "the quantity of water to be considered is the minimum rate of flow during a drought with an average 50 year frequency as obtained by a licensed Professional Engineer, Hydrologist, Geologist or similar qualified person."

Rural Fire Protection Districts have experienced difficulty in obtaining written certification that impounded water supplies or flow streams that established impounded water supplies or flowing stream supplies meet the minimum ISO criteria of 30,000 gallons under drought conditions as specified above. The problem is twofold: 1) it appears that many professional civil engineers lack the background and experience to make such certifications. Geologist and hydrologist are more suitable for certifying natural water supplies. 2) Professionals typically charge from $300 a day and up to research natural water sources and provide professional certification; typically rural fire protection districts simply do not have this kind of money available for this purpose.

It is generally recommended that local fire protection jurisdictions contact the State or county office of soil and water conservation as an important resource in establishing alternative water supply sites, obtaining guidance on installing dry type fire hydrants at specific water supply sites, and obtaining funds for installing dry hydrants and certifying alternative water sources. Other resources for assistance resources that can be consulted for assistance on certifying ground level water supplies include

- NRCS—Natural Resources Conservation Services;
- DEC—Department of Environmental Conservation; and
- Army Corp of Engineers.

4. Developed sources of alternative water supplies.

There appears to be a growing trend in suburban and rural areas to avoid the complications and possible unreliability of static and flowing water sources by using developed sources of water supplies to meet water supply needs for fire protection. This involves providing containers for water supplies at strategic locations throughout a fire district. The storage supplies are easy to calculate, these can be readily assessed by ISO field representatives, and in the long run maybe more economically feasible. Most important reliability of each water supply is constant because the supply source is not subject to drought or their environmental conditions. Furthermore, fire protection jurisdictions do not need to have the water supply certified by a professional person. The following approaches for providing alternative water supplies can be directly evaluated by a representative of the ISO without a third party certification.

- Cisterns: Cisterns are typically installed below ground water storage tanks constructed of reinforced concrete equipped with a dry hydrant. **Figure 6-1** shows a cutaway view of a cistern documented in NFPA 1142. (__) This same reference provides guides to cistern specifications, construction protection against freezing, inspection, maintenance, and topping off the water supply.

- Fiberglass underground storage tanks: The Xerxes Corporation manufactures fiberglass reinforced plastic (FRP) tanks that offer an important option for the underground storage of water. The features and benefits include

  - rustproof, long lasting fiberglass;

  - internal rib design for added strength and robustness;

  - ease of shipment and installation;

  - manufactured to applicable requirements of Underwriters Laboratories (UL) 1316 including watertight specifications to prevent leakage or seepage into the container;

  - comparable with American National Standards Institute/National Science Foundation (ANSI/NSF) Standard 61;

  - available in standard and custom designs;

  - available in single-wall, double-wall, and triple-wall construction to meet local soil conditions;

  - available in holding capacities from 600 gallons to 50,000 gallons; two or more tanks may be interconnected to provide large capacities; and

  - a special 30,000-gallon size is manufactured specifically for fire protection requirements with a full bottom antivortex outlet that can be concreted to a dry hydrant as illustrated in **Figure 6-2**.

- Aboveground storage tanks: Fiberglass and steel storage tanks maybe mounted aboveground as a water storage supply for fire protection. However, this type of storage is limited to regions of the country where temperatures are above freezing; the cost of auxiliary heating in other climates is probably unacceptable. In Southern and Western States, railroad tank cars located on a track siding are kept full of water for fire protection. A 6-inch supply line goes from a suction sump in the tank to a nearby wet-type fire hydrant.

- Elevated water storage tanks: Elevated tanks maybe installed as a sole source of water supply for fire protection for rural fire departments. These tanks typically are supplied by a well pump and are located at fire stations for refilling fire apparatus water tanks, including mobile water tankers.

- Driven wells: These types of wells generally are installed in rural areas for crop irrigation and may be used for alternative water supplies under specified conditions. The well supply may have a feed main to a dry hydrant located by an access road. This permits testing the well supply for flow rate capability. The volume of water from a given well needs to be determined by draw down tests.

- Swimming pools: The ISO *only accepts* swimming pools as an alternative water supply source where:
  - the pool capacity equals or exceeds 30,000 gallons of water;
  - the pool remains at the full level year round;
  - access is provided to the pool for a mobile pumper to draft water from the pool, **or** a 6-inch pipe is run from the pool to a pumper connection location point.

  Each individual swimming pool needs to be evaluated by an ISO Field Representative.

- Private domestic water systems: Suburban residential communities and large farmstead properties, especially dairy farms with milking parlors, actually have small self-contained water systems supplied by wells that have the capability of delivering 250 gpm or more. It is determined from tests that a dedicated water system can supply 250 gpm for 2 hours, the property association or the property owner should contact their insurance carrier(s) to initiate an ISO evaluation of the system.

- Municipal water supply systems: Recognized fire hydrants on municipal water systems (i.e., Those hydrant tested to demonstrate a delivery of 250 gpm or more for 2 hours) may be used to fill mobile water tankers or to relay water to protect areas beyond 1,000 feet of the hydrant. This is a very **important** source of water supply when planning alternative water delivery program, since it may be one of the most reliable water sources for fire protection beyond water mains.

5. **Dry fire hydrant programs.**

Alternative water supply program for suburban and rural areas need to consider *carefully* the advantages of installing dry (fire) hydrants at ponds, lakes, river, streams, or on water storage tanks to provide ready access to water supplies for sustained structural fire suppression capability. The term "dry hydrant" refers to a special type of hydrant that is installed at or on a water source that has **no** static water head or pressure and is intended to provide water *strictly* for fire protection. Dry hydrants significantly improve structural fire protection in suburban and rural areas where there is no municipal water supply for the following reasons:

- Each individual dry hydrant has a specific location at a named water source to maximize the water available for supplying a fire department pumper.

- A fire department pumper crew can connect one section of hard suction hose rapidly between the dry hydrant outlet and a mobile pumper intake of the correct diameter. The eliminates the time-consuming effort to connect two or more sections of suction hose together, attach a straining device, and correctly placed the suction hose and strainer in the water source.

- A solid earthen, blacktop, or concrete pad is prepared near the dry hydrant for safety of the fire apparatus and personnel working at the dry hydrant location.

- Locate dry hydrant cap suction connections within 18 feet of a hard surface road or drive-off to the pumper location pad. In snow-belt areas, the drive-off pad area needs to be arranged for easy snow plowing. There should be no significant accumulation of snow around the dry hydrant area during and after snowfalls.

- All dry hydrants used for providing water to an ISO *Graded Area* need an individual hydrant identification number posted near the dry hydrant. Dry hydrant locations and the reference number need to be plotted on a fire protection jurisdiction highway map. This map will improve the ability of ISO personnel to evaluate water delivery capability throughout a fire protection district.

Prior to the installation of a dry hydrant, it is essential that the location of each dry hydrant be carefully planned, design criteria established, construction details documented and prepared but **not** signed. This process is necessary to obtain installation permits from the authority having jurisdiction (AHJ). Suburban and rural areas are encouraged to contact the county office of soil and water conservation, the regional office of natural resources conservation (NRC) and/or the department of environmental conservation (DEC). The NFPA's Standards Committee on Suburban and Rural Fire Protection is working with DEC to provide a means whereby the agency will supply installation drawings for a fire hydrant at a specific water site based on drought conditions in the area and the potential reduced capacity from freezing ice over the intake port. It is then expected that the ISO will accept the dry hydrant installed according to the drawings without a further certification process. However, such a program is still a way from reality.

Figure 6-3 illustrates a typical dry hydrant installation for a water source that has been certified for hold 30,000 gallons of water during drought conditions. It is essential to observe the following when examining and evaluating dry hydrant installation drawings and the actual field installation:

- Note that a hard surface or *packed dirt pad* needs to be provided for locating a supply pumper or a fill site pumper at the dry hydrant location.

- The pad area and surroundings need to be cleared of encroaching ground cover and maintained to prevent soil erosion.

- A dry hydrant marker with the dry hydrant number needs to be placed in a   conspicuous location near the *pad* area.

- Figure 1 indicates a measure of 20 feet of elevation difference as a maximum figure for drafting water between the indicated surface level of the water and the centerline of the "steamer" connection or hydrant outlet. Both the mechanical efficiency of the mobile pumping unit and the elevation above sea level act to affect the ability to draft water with an elevation differential of 20 feet. A standard 10 feet of head differential between the two elevation points referenced should be established where possible.

- The distance between the raked-off finished ground surface and the steamer hose connection outlet centerline should be 12 inches. It is essential that the centerline of the hydrant hose connection be below the centerline of the pump intake in order to avoid an air lock in the suction hose.

- The dry hydrant riser pipe, top elbow, and outlet connection (i.e., hydrant head) are typically 6 inches in diameter. The threaded hose connection provides for a 6 inch waterway. The hose threads should typically be National Standards Thread (NST). Hose adapters **may need to be provided** for any pumping units that still do not have NST hose threads.

[Sidebar: The issue of proper hose threads is extremely important. In 2003, the USFA reported the identification of 19 difference "steamer" or large outlet diameter hose threads in use in the United States.]

- Low-level vegetation, preferably one that can be mowed, should be planted around the riser pipe and along the embankment to prevent erosion

- The underground lateral pipe connecting the strainer with the elbow section or single elbow to the riser is usually 6 inch in rural areas. A larger size lateral pipe maybe needed where:
  - A dry hydrant flow greater than 1,000 gpm is needed; or
  - The length of the lateral suction pipe exceeds 50 feet.

- In either case, an increased pipe size may be needed to reduce friction loss at higher flows. The following guideline is useful to estimate maximum flows in PVC pipe per 100 feet of pipe with connector fittings:

Pipe Diameter	Flow in gpm
6 inch	1,000
8 inch	1,250
10 inch	1,500

It cannot be reinforced enough that the water level above the installed strainer needs to be 2 feet above the minimum drought level certified for a specific water source to avoid a vortex at the entrance to the suction pipe and loss of prime on the intake side of the mobile pump. A vortex box to house the pipe strainer maybe constructed to avoid this problem.

6. Methods of delivering Alternative water supplies.

The ISO does not prescribe any specific method of delivering alternative water supplies to a fire site. In fact, ISO encourages the development of innovative techniques and methods to improve current approaches to water delivery operations for fire protection beyond 1,000 feet of recognized fire hydrants. It appears that the majority of existing water delivery program have evolved from one or both of the following approaches to transporting water from a recognized source location (i.e., pond, lake stream, river, storage tank, etc,. as noted above) to an identified fire risk site.

- Use of mobile tankers to "haul" or "shuttle" water between a designated water fill site and the fire site.

- Use three or more mobile pumper to transport water through one or more hose lines from a water source to the fire site. There are two basic types of water relays: 1) closed-end relay, and 2) open-end relay.

Establishing an Alternative Water Delivery Program

The first organized effort to establish alternative water supply programs appears to have emerged in the late 1970s. There is documentation of these programs in the States of California, Georgia, Florida, New York, and North Carolina. One specific program was established between the ISO and the Wyoming County Bureau of Fire in western New York State. The State Fire Coordinator for the Bureau of Fire was interested in developing a program that would improve the semiprotected areas of the county to a protected classification for insurance rating purposes. It was reasoned that not only would this effort improve rural fire protection but it also would reduce property insurance rates rather significantly. Working with the ISO Field Representative out of Rochester, New York, deputy fire coordinators started a Master Plan to deliver needed fire flows at representative sites throughout the individual fire districts in the county. There was little published material beyond the general criteria obtain from the ISO.

Wyoming County was, at that time, a mostly rural county with several small communities. Dairy farming was the primary livelihood. It was estimated that the county had around 40,000 people and 400,000 thousand cows; it was rural America in every sense of the word. However it was a thriving county and energized to do things better.

The county was very fortunate to have a County Office of Soil and Water Conservation, with a graduate of Cornell University in both Agricultural Engineering and Civil Engineering. The individual was essential to the certification of farm ponds, small lakes, and streams running through the county that could be developed as water resources for the alternative water delivery program. By the mid-1980s the program was in progress with seven participating fire districts. However, the progress was slow and tedious because of all the paperwork that needed to be generated for conducting a water delivery demonstration for the ISO.

The small village of Strykersville, New York, in Wyoming County is certainly one of the first ISO *Graded Areas* an improved Public Protection Classification (PPC™) based on a successful water delivery demonstration using both a tanker shuttle operation and a water relay from a nearby creek to delivery 1,550 gallons of water a minute at the Catholic Church near the center of the community. The PPC Class actually improved from a Semi-Protection Class 9, since there was no recognized water supply to a Class 7 Protected Classification. This occurred in 1990 and it was estimated at that time that the residents of the community would receive an insurance premium reduction of $1.37 per $100 of evaluation; commercial property would be more, but this was not determined for each property in this class for the community. Voluntary contributions to the volunteer fire department tripled the next year in gratitude for this achievement. Approximately 35 percent of the county fire districts now have an ISO PPC of 8 or better. This process continues in the county, but it has slowed down because of the lack of funds at the county level to continue this effort.

A significant program was initiated in 2000 to give alternative water supply programs a national perspective and to encourage suburban and rural fire protection districts without recognized water supplies to plan and implement water delivery programs. The USFA joined in a partnership effort with the ISO to develop A User Planning Guide and a User Workbook on Alternative Water Supplies. In addition, two companion videotapes are available with the titles 1) *Planning for Alternative Water Supply* and 2) *Implementing An Alternative Water Supply Program*. These videotapes were introduced at the International Association of Fire Chiefs (IAFC) Conference, Denver, Colorado, in August of 2005.

[Sidebar: The references at the end of this chapter provide information on how to obtain the *User Plan Guide*, the *Workbook*, and the two videotapes on alternative water supplies from the USFA.]

The USFA encourages fire departments in conjunction with the communities they serve to develop and implement long-range plans. The USFA has developed a planning model that incorporates the systems approach to planning on public fire defenses. This planning model is easily adapted to establish an alternative water supply project for a specific fire protection jurisdiction as follows:

Phase I: Preparing and organizing to plan.

1) A fire protection jurisdiction recognizes the **need** to deliver water in sufficient quantities to control and extinguish structural fires in response areas without water mains supplying fire hydrants.

2) An organization meeting is held with at least the following representatives attending:

   a) The ranking political official (i.e., mayor or township manager) at the local government level and the county executive at the county level of government.

   b) The elected official responsible for fire protection.

   c) The fire chief and fire company officers for the fire district under consideration, and county fire officials where the county has planning responsibilities.

   d) A community water supply superintendent if there is a recognized water supply system in the fire district.

e) The ISO Field Representative who has responsibility for the identified fire protection jurisdiction.

3) The ISO Field Representative discusses the ISO criteria for obtaining a recognized alternative water supply program for the graded area the process to reach this goal.

[Side bar: The ISO Field Representative plays a **key** role in all three phases of the alternative water supply planning process. Both the *User Guide* and *Workbook* are based on a water delivery program that will result in the *Graded Area* receiving an ISO PPC.]

Phase II: Conduct the Planning Process

1) Identify resources for conducting alternative water supply delivery demonstration project. These resources include a) mobile fire apparatus and equipment and response personnel, and b) available outside aid in the form of mutual-aid and automatic-aid from surrounding fire districts.

2) The ISO Field Representative identifies a minimum of five representative fire risks in the *Graded Area* for determining a response evaluation. In cooperation with local authorities, the ISO Field Representative selects **one** representative fire risk site where a water delivery demonstration is to be conducted.

3) Arrangements are made for the ISO Field Representative to witness a water delivery demonstration at the selected fire risk site.

4) After a successful water delivery demonstration has been completed, the ISO Field Representative will evaluate water delivery to the remaining fires risks in the *Graded Area* using a computer simulation model developed by the ISO.

[Sidebar: ISO's Alterative Water Supply Computer Simulation Model reduces the time of preparation required by a fire district before a Grading Reclassification can be completed. This overcomes a major problem associated with the early work on alternative water supplies as identified above.]

Phase III: Adopt and implement the plan.

1) After the ISO completes the alternative water delivery capability evaluation throughout the *Graded Area*, the fire protection jurisdiction submits a request to ISO, Inc. for a Public Protection Reclassification.

2) The reclassification includes a current evaluation of: a) receiving and handling fire alarms, b) fire department, and c) water supply. Note: Results of the alternative water supply evaluation replaces a component evaluation of a municipal water supply.

3) The alternative water supply program is implemented in the fire protection jurisdiction. This means that all future alarms of fire for a structure fire will receive a response of equipment and personnel to place in service an alternative water supply to meet as close as possible the needed fire flow for the fire risk.

Summary

The model alternative water supply program follows the steps below:

◆ Establish needed fire flows throughout the *Graded Area*.

◆ Define the geographical area to be protected.

◆ Identify the representative fire risks and basic fire department response information to each potential fire site.

◆ Identify the water sources that will be used to transport water to the representative fire risks, and review the certification of each water source to determine that it is capable of delivering 250 gpm for a during of 2 hours under the specified drought conditions.

- Determine the water transport methods to be used for each representative fire site. This is typically a water shuttle operation using mobile tanker apparatus that maybe be augmented by a water relay operation.
- Identify fire apparatus resource capability.
- Evaluate fire apparatus travel plans.
- Prepare a preliminary operations plan.
- Conduct a water delivery demonstration project.
- Analyze the completed water delivery demonstration.
- Prepare operation plans for the remaining representative fire site.
- Request an ISO Reclassification.
- Determine the benefits of the improved PPC on Property Fire Insurance Rates.

CHAPTER 7 IMPACTS OF FIRE FLOW ON DISTRIBUTION SYSTEMS WATER QUALITY, DESIGN, AND OPERATION

Background Statement

In 2002, The American Water Works Research Foundation and KIWA of the Netherlands convened a panel of engineers, scientists, university professors, municipal water works officials, senior fire officers, and insurance industry personnel to study, explore, and evaluate, as this Chapter title indicates, *Impacts of Fire Flow on Distribution Systems Water Quality, Design, and Operations*. This study was jointly funded by the AWWA, the AWWA Research Foundation, and KIWA of the Netherlands, who assume no responsibility for the content of the research study reported in a publication by the Chapter title above or the facts and opinions of the panel participant's as expressed in the report.

The following information is excerpted from this study because the information presented is essential to understanding current issues associated with public-sector water supplies and the potential impact that the private sector has on needed water supplies for fire protection. It should be understood that there are two approaches to the confinement, control, and extinguishment of developing fires in structures. The conventional approach is to have a public fire department respond to a structure fire with engine companies and ladder companies to "handle" everything from a wastebasket fire to a fully involved structure fire. The alternative approach is to install automatic sprinkler systems in both residential and commercial properties either to extinguish a developing fire immediately, or to confine a fire to the room of origin to be completely extinguished by a fire department.

The installation of automatic sprinkler systems transfers a significant portion of the cost of water supplies for fire protection to the private sector. The private sector assumes the cost of the installation of automatic sprinkler systems, maintenance, and periodic testing. There are economic incentives in the form of insurance premium reductions for commercial property owners with installed and maintained sprinkler systems and even some insurance incentives for residential sprinkler system installations.

As stated in previous chapters, the primary objective of a public water system is to provide sufficient potable water at an acceptable level of water quality now determined by the EPA in the United States. The decision for a public water supply to provide fire flows can have significant impacts on the design and operations of the systems. This is why there is a large number of small villages and towns under approximately 5,000 population that do not have installed fire hydrants on a small water distribution system.

Typically, the election to provide fire flows and fire hydrants results in increased water supply pipe diameters, leading to higher capital costs, and greater provision for reliability and redundancy in the distribution system. It may also, however, have some negative water quality implications. This "oversizing" to meet what some consider to be relatively infrequent fire events can result in increased water resident times in larger size pipe, thus increasing the possibility of disinfectant residual loss, and enhancing the formation of disinfection byproducts, and bacterial growth in the water mains. Larger diameter pipes also result in lower water flow velocities in the water system that lead, in turn, to the deposition of sediments.

Background

The study group determined that there is no legal requirement in the United States for the water distribution system to provide fire flows. Municipalities and township may develop ordinances that require water systems to provide fire protection. At the same time, an inadequate fire protection system provides a false sense of security. Fire flow requirements generally are based on fire codes developed by independent groups including the Building Officials and Code Administrators (BOCA), the International Conference of Building Officials (ICBO), the Southern Building Code Congress International (SBCCI), and the National Board of Fire Underwriters' up to 1970. Since 1970, the insurance criteria has been developed by the ISO.

In the absence of a fire, the obligation to provide fire protection substantially extends the time that the water, after treatment, resides in the distribution system, including its service reservoirs, before it reaches the users. Degradation of water quality in conventional potable water distribution systems has been shown to be a function of the length of time that water is retained in the distribution system and the very slow velocities of water in the lines that feed residential areas, to the point where they protect microorganisms in the water and, over time, restrict flow capacities of the pipes. (Geldreich 1996)

Water quality changes that occur within the distribution system are of increasing concern.

Water distribution system design often is dictated by the need to provide fire flows. This leads to oversizing of the system for most normal conditions. An oversized system increases the resident time of water, increasing the possibility for depletion of disinfectant residual and the formation of disinfection by-products. An oversized water distribution system also increases the associated capital and operation and maintenance (O&M) costs. This is particularly true in the case of pumping, when elevated tanks are maintained at high levels in order to provide sufficient fire flow capacity and minimum pressures on fire hydrants. Therefore, there is a clear need to evaluate fire flow requirements and to determine improvements in water quality and reductions in capital and O&M costs that could be achieved if fire flow requirements are reduced or eliminated.

Study Objectives and Scope

The objectives of the stated project were as follows:

- determine the economic and water quality impacts of designing systems to meet fire flow requirements;
- review national and international criteria for providing fire flows; and
- identify alternative methods and technologies for firefighting appropriate for present-day situations.

This review of the study report addresses each of these topics in order.

The Impact on Water Quality by Designing Water Distribution Systems to Meet Fire Flow Requirements

In order to evaluate the economic and water quality impacts associated with fire flow requirements, hydraulic water quality modeling of hypothetical and actual distribution systems was conducted. The EPA's EPANET model was used to conduct the modeling. The hydraulic evaluation considered changes in infrastructure sizing and operating practices, primarily related to storage volumes, which could be realized if fire flow requirements were reduced or eliminated. The economic impact of these changes was estimated to evaluate the financial impacts of providing fire flows. The key water quality parameter considered was water age. The residence time of water in the distribution system can be a key indicator of water quality, as it plays a major role in determining disinfectant residuals and disinfection byproduct formation. The significant findings are as follows.

Mathematical models were applied to various systems designs and different levels of fire flow in order to study the resulting economic and water quality impacts. The following overall approach was used in conducting this analysis:

1) A modified version of EPANET distribution system model (Rossman 2000) was applied in steady-state mode to a specified network design to determine if it would meet minimum pressure requirements under a) maximum day demand plus fire flow Requirements, and b) peak hour demand conditions.

2) Three levels of fire protection were studied

   a) The first level is maximum day demand plus fire flow capacity in which fire flows meet the needed fire flow requirements as defined by the ISO for Public Protection Requirements (see Chapter 5 for details on the ISO needed fire flow calculations).

   b) The second level is maximum day demand plus reduced fire flow capacity associated with the hose stream requirements that are needed to augment fire protection provided by automatic sprinkler systems.

   c) The third level is no fire flow requirement (i.e., peak hour demand for consumer consumption). This alternative would be consistent with a dual water system briefly discussed under alternative technologies later in this review and in Chapter 8 of the manual.

3) A modified EPANET model was applied using an interactive approach in which pipe diameters were downsized under the constraint that minimum pressure requirements were met under both peak hour conditions and maximum day plus fire flow requirements at all demand nodes. This step was repeated for the three levels of fire protection described in the previous step.

4) The EPANET model was applied in extended period simulation (EPS) mode to determine distribution system storage requirements and to ascertain that minimum pressures and flows could be met over the full duration of a fire (i.e., 2 to 4 hours).

5) EPANET model was applied in the EPS mode to estimate water age under minimum daily demand conditions in order to calculate the average demand-weighted water age.

6) The results of the EPANET minimum day demand application were examined to determine whether sufficiently high velocities occurred to ensure "self-cleaning" of the pipes. The criterion used by KIWA that a minimum velocity (1.3 feet per second(fps)) occurs at least once a day was employed as the evaluation measure.

7) A cost model developed by EPA (Gumerman, Burris, and Burris, 1992; Clark et al., 2001) was applied to determine the capital costs of pipes, tanks, and other facilities for the various distribution system designs associated with each level of fire protection facilities.

8) The redundancy and reliability of each of the distribution system designs was qualitatively evaluated.

In addition to the standard procedures outlined above, a tank water age model, CompTank, developed as part of an earlier AWWA Research Foundation sponsored project (Grayman et al., 2000) was applied to develop general guidelines for storage requirements to meet various levels of fire protection.

EPANET Distribution System Model

Hydraulic analysis of flows and pressures in a distribution system has been a standard form of engineering analysis since its development by Dr. Hardy Cross in 1936. Water distribution system computer models have been in use since the middle 1960s and have evolved into sophisticated, user-friendly tools that are capable of simulating large distribution systems (Walski, Chase, and Savic, 2001). In more recent years, the ability to model water utility and water age has been added to hydraulic models (Clark and Grayman, 1998). There are many commercial models that offer a wide range of capabilities in distribution system modeling EPANET is an open-structured, public domain hydraulic and water quality model developed by USEPA and is used worldwide (Rossman, 2000.) EPANET was selected for the stated research project.

In order to facilitate the examination of required pipe sizes, the standard EPANET model was modified for use in the study project. This modified version allows the user to examine all nodes quickly in terms of their ability to deliver specified fire flow quantities while meeting a specified pressure requirement at the fire flow node; 20 psi was selected. The model also may be used to automatically downsize pipes to the minimum diameter required to deliver a specified fire flow plus normal usage.

Costing Model

EPA has developed a model consisting of cost equations for pumps, pump stations, including new facilities and expansion of existing pipes (Clark et al., 2001). The model represents the base construction cost data for the purchase and installation of an item, such as a particular type of pump installation. Cost estimates can be developed based on a series of separate items that can be added to the base installed costs. To convert the base construction cost into capital cost, the cost data must be aggregated for the entire project, and additions made that include the following items:

- general contractors overhead and profit;
- engineering;
- land or right-of-way acquisitions;
- legal, fiscal, and administrative costs; and
- interest during construction.

O&M requirements for pumping stations include electrical energy, maintenance materials, and labor. Total O&M cost is a composite of the energy, maintenance material, and labor costs.

Observations from Model Runs

💧 As the development increases consumer demand, peak demand flow becomes a controlling factor and pipe size is determined by peak flow rather than maximum demand **day** plus fire flow requirements. When peak flow is the controlling factor, then the system also should meet fire flow requirements. As peak flow decreases, water demand decreases; a system design based on peak flow will not necessarily meet fire flow requirements.

💧 A network design using a looped feeder system results in overall smaller diameter pipe sizes and lower water usage than either the regular grid system or the branching network system. However, the branching network results in significantly lower costs than the other networks because of the reduction in total length of pipe despite the larger pipe diameters required. This may explain why many distribution subdivisions are built with branching type systems rather than looping systems. It was further identified that the nonlooped configuration used in this analysis results in longer connection lines, as compared to the looped configurations.

💧 Examination of the branching systems shows that such systems provide no service redundancy and are thus inherently less reliable (i.e., more vulnerable to service shutdowns) than looping systems. In a branching system, any outage along a single pipe results in loss of service to all downstream consumers. In a looping system, alternative paths generally will provide some degree of service even if one pipe is out of service.

💧 As pipe size increase to meet fire flow requirements, the average maximum velocities in individual pipes tend to decrease, resulting in a greater potential for sedimentation, deposition, and water age. This conclusion is supported by the larger percentage of pipes that do not meet the KIWA velocity goal (minimum velocity of at least 1.3 fps at least once a day as the fire flow requirement increases).

Analysis of Distribution System Storage

Storage is an important component in almost all distribution systems. It can serve the following purposes (Walski, 2000):

💧 Equalization: Demands in a water system generally vary over the course of the day, while water utilities prefer to operate their treatment facilities at a relatively constant rate. Distribution system water tanks and reservoirs frequently operate in fill and draw operations over the course of the day, thus providing the storage to accommodate the constant variations in water supply and demand. However, time-of-day variations in energy pricing may influence when wells and pumps are operated.

💧 Pressure maintenance: The water level in tanks and reservoirs largely determine the pressure in areas served by a storage facility. In order to provide sufficient pressure, in many situations, particularly at the top of the pressure zone, the water level in a storage facility must be maintained within a specific range.

💧 Fire storage: Required or needed fire flows can be provided through a combination of fire storage in water tanks and reservoirs, or through larger transmission lines and increased treatment capacities. In many water systems it appears that fire storage is the more economical means of meeting fire flow requirements. Dedicated water in storage for fire protection needs to be recycled weekly to prevent excessive aging and sedimentation. This can be done by bringing the tank "on line" and refilling from the top using a total flow meter to replenish the gallons of water drained from the tank. This technique keeps the tank 90 percent full at all times.

💧 Emergency storage: In addition to fires, other emergencies such as power outages, equipment failures, water main breaks, and temporary loss of water supply facilities can result in insufficient water supply within the distribution system. Storage provides a mechanism for providing water under such emergency conditions.

In order to accommodate the various purposes of a water tank or reservoir, they are generally designed with a given amount of capacity targeted to each of the following specific uses:

💧 Equalization storage: Storage used to allow for normal fill and draw patterns.

💧 Ineffective (passive) storage: Storage used to provide the minimum pressure requirements.

💧 Fire storage: Storage to meet sufficient, required, or needed fire flows.

💧 Emergency storage: Storage reserved for emergencies other than fires.

The actual storage requirements in each of these categories vary in different storage facilities based on local regulations, operating conditions, and hydraulic conditions. It should be understood that at some water utilities, fire storage and emergency storage may be combined, based on the assumption that both fires and other emergencies are rare events and the simultaneous occurrence of fires and other emergencies are unlikely.

[Sidebar: This study was completed before 9-11 at the World Trade Center. Today, FEMA considers the probability threat level for simultaneous events, which could way overtax even a very large city water supply.]

The primary emphasis of the study under review is to evaluate the impact of fire flow requirements on water storage capacity. Storage capacity is aggregated into three categories: 1) equalization, 2) fire demand, and 3) reserve storage. Reserve storage corresponds to capacity that is not used for either a joint equalization or simply fire flow demand and includes both ineffective storage and emergency (other than fire) storage. In order to develop a general relationship, the following assumptions were made

💧 The tank or reservoir is completely and instantaneously mixed during the fill cycle.

💧 The storage facility operates with a 12-hour draw and 12-hour fill cycle at constant fill and draw rate with the water level variation over the full range of the equalization storage volume.

💧 The water level in the tank is at its maximum level at the start of the draw cycle and again at the end of the fill cycle.

The types of storage and the fill and draw pattern used in the analysis are illustrated in **Figure 7-1.**

Though primarily designed for the hydraulic purposes outlined above, storage also can have a negative impact on water quality due to age as previously identified. Loss of disinfectant residual, bacterial regrowth, taste and odor production, and formation of disinfectant by-products are potential water quality problems associated with water aging. Improper mixing in storage facilities can exacerbate the water aging problems by creating dead or stagnant zones of even older water (Grayman et al., 2000.) A quantitative assessment of the impacts of providing fire storage on water age is provided in **Figure 7-1**.

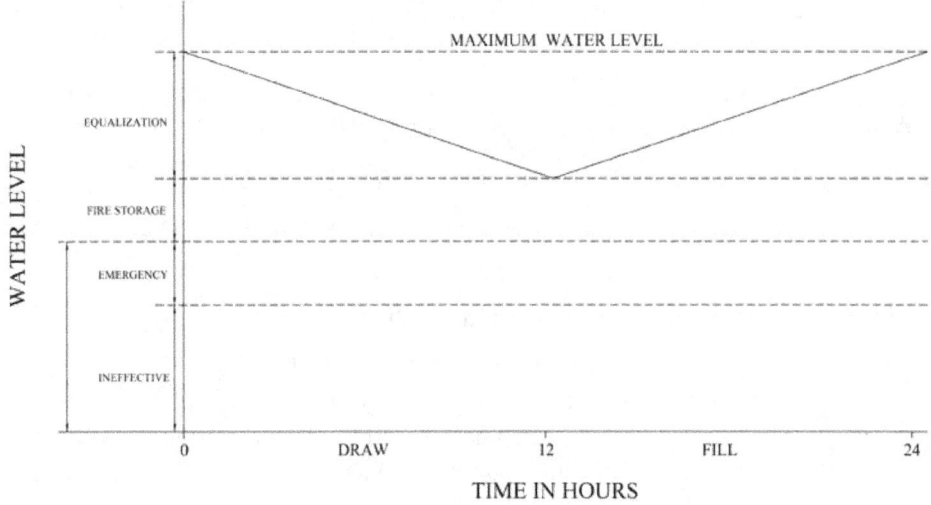

Figure 7-1

Factors Used in the Analysis of Water Storage

Study Set 1: Summary and Conclusions

Designing a water system with sufficient capacity to meet fire flow requirements or needed fire flows can result in major economic and water quality impacts. Typically these impacts are manifested in increased pipe diameters, greater provision for reliability and redundancy in system design, increased system costs, and increased potential for reduced water quality.

A method has been developed for approximating minimum pipe diameters required in a water system for various levels of fire flow and peak hour conditions. The economic and water quality consequences of these design choices have been evaluated.

The methodology has been demonstrated by applying it to a realistic, though hypothetical distribution system. System requirements for four different scenarios were evaluated:

1) Existing pipe and storage sizes.
2) Modified (downsized) pipes and storage that will still meet the full fire flow requirements.
3) Modified pipe sizes and system storage to meet a reduced fire flow requirement associated with having sprinkler systems in all structures.
4) Modified pipe sizes and system storage with no fire flow requirements.

The adequacy of the various system design options was evaluated in terms of their capability to meet fire flow requirements and to satisfy normal demands. Each option was examined in terms of its ability to meet maximum day conditions with a fire flow requirement superimposed and to meet peak hour conditions. The analysis includes both a steady-state assessment and an EPS that tests the ability of the system to deliver flow over the duration of a fire ranging from 2 hours to 4 hours. Subsequent to the identification of the system configurations under the four design options described above, additional modeling was performed to:

1) Determine the cost of the resulting networks.

2) Determine the water quality consequences of the reduced pipe diameters and storage requirements.

In order to make the results more generally applicable, the same method was applied to a generalized network using four different design criteria, and three different fire flow requirements, under three different population densities. A series of observations result from this analysis concerning the impact of these assumptions on pipe diameters, pipe costs, water age, and velocities.

Review of National and International Criteria and Providing Fire Flows

Water companies throughout North America and Europe, whether private or publicly owned, operate within a regulated environment and strive to provide the highest levels of service to their customers at levels of costs that reflect good value for the money invested in the water treatment and distributions systems. (Water UK and LGA, 1998) In terms of water supply for domestic purpose, "the highest level of service" usually means to provide a water service to customers at acceptable minimum flow and pressure, as well as supply water that meets applicable water quality regulations and is aesthetically pleasing. Water requirements for fire suppression or installed fire protection, on the other hand, are usually added to the requirements for domestic purposes and are commonly calculated according to discrete areas belonging to different fire risk categories that are based on type of construction and building size. As discussed in greater detail below, fire flow requirements vary greatly among countries. These variations may be attributed to the *differences in construction materials and structural types* that exist among countries. The following review is by country that provided literature on water supply requirements.

United States

Throughout the United States, it generally is left up to the local authorities to determine the fire flow requirements for their jurisdiction. The fire flow requirements sometimes are specified in the model building codes or fire codes; however, if the political jurisdiction wishes to determine its own set of fire flow requirements, there are three methods reported in the literature from the AWWA. Each of these methods is identified below without full elaboration, for two of the methods are discussed in other referenced Chapters, and a third method has not been adopted by any standards or codes. Basic details are provided so that these methods can be compared and contrasted to other countries.

🔹 The ISO Needed Fire Flow Method: Refer to Chapter 6. The ISO Method is used to determine needed fire flows for insurance rating purposes. This is a recommended practice for communities but there is no legal requirement that political jurisdiction meet needed fire flow criteria. However, municipal water supplies that do meet needed fire flow criteria may find that their commercial property owners are paying higher insurance rates; it may have less impact on residential property owners. The ISO method for calculating the needed fire flow considers the construction (C_i), occupancy (O_i) exposure (X_i), and communications (P_i) factors of a building or fire division. Exposure and communication factors reflect the influence of exposed and communicating buildings on the needed fire flow (ISO

2001). The ISO Method is most likely to produce the most realistic requirements (Farrell, 1996). Fire flow is calculated by the formula: $NFF_i = (C_i)(O_i)(X + P)I$.

- Illinois Institute of Technology Research Institute (IITRI). The IITRI method is based on data collected from actual fire in server types of commercial occupancies in Chicago and to determine the water application rate necessary to control a fire as a function of the fire area. Using a curve fitting analysis, the IITRI method developed two equations solely based on flow area. The method has not been adopted by any code- or standards-making organization.

- Iowa State University Method. The ISU method for calculating a required fire flow addresses both the quantity of water required to extinguish a fire, and the effects of various application rate and firefighting techniques. The equation used for this method is based on the consumption of fuel being dependent on the available oxygen supply in a 90 percent closed compartment or space and vaporization of applied water into steam. (AWWA, 1998) A limitation of this equation is a result of the assumption that the entire compartment or space be entirely involved in fire. Therefore, the critical consideration involved in applying the ISU method is how a fire department fights structural fires. (Burns and Phelps, 1994) This method is discussed further in Chapter 6, and it should be noted that this method is fully explained in NFPA 1142.

In addition to these basic methods, building codes and standards in the United States also specify water supply requirements for fire protection. The following building code information was identified from a review of the literature.

BOCA: The BOCA National Fire Code references the NFPA Codes and Standards for the determination of required fire flows.

ICBO: The Uniform Fire Code defines fire flow as the flow rate of a water supply, measured at 20 psi residual pressure, that is available for firefighting (ICBO, 1997). For one- and two-family dwellings, a fire area that does not exceed 3,600 square feet, the minimum fire flow requirements are 1,000 gpm. For other types of construction, the minimum required fire flow and flow duration for buildings are indicated in **Table 7-1**.

Table 7-1: Minimum Required Fire Flows and Flow Duration

Fire Area in Sq. Ft.*	Fire Flow in gpm	Duration in Hours
3,600 — 70,900	1,500—2,750	2
13,400 — 128,700	3,000—3,750	3
23,300 — 295,900 and greater	4,000—8,000	4

Source: ICBO 1997 *Fire areas for specific types of construction are provided in Appendix III-C, Fire Hydrant Locations and Distribution of the UFC (ICBO) 1997.

NFPA: NFPA leaves it up to the Authority Having Jurisdiction to determine the required fire flows for the local conditions in the form of local codes, ordinances, or resolutions. These local codes may, however, specify the minimum water supply requirements that must be available for firefighting purposes in areas where adequate and reliable water supply systems for firefighting purposes do not exist. The fire department having jurisdiction should perform an onsite survey of all building, including types of construction, occupancies, and exposures within the applicable jurisdiction to obtain the information needed to compute the minimum water supplies required (NFPA, 1999). NFPA 1142 provides calculations for determining the minimum firefighting water supply for structures based on their exposure hazard, total area of the structure, occupancy hazard classification number, and the construction classification number. Tables specify the minimum water supply to be 2,000 gallons and the maximum supply of 5,000,000 gallons that must be available in storage for firefighting purposes.

Canada

It is the responsibility of the provinces and territories throughout Canada to determine fire flow requirements. The following is a summary regarding some of the codes and guidelines used by local authorities in Canada for determining the required fire flows for their province or territory.

- National Building Code (NBC): The NBC does not dictate the required fire flow, but states that an adequate water supply for firefighting should be provided for every building (Farrell, 1996).

- Alberta Building Code: In 1973, Alberta became the first province to adopt the NBC by reference, making some amendment as noted in the Alberta Code (ABC) (Farrell, 1996). The ABC provides a formula for calculating the supply of water to be available for fire-fighting purposes. The formula is based on the total building volume, a water supply coefficient, and a spatial coefficient. The water supply coefficient is selected from a table provided in the ABC, while the spatial coefficient is equal to either 1.5 or 1.0 based on the limiting distance between buildings. The water supply should be capable of being delivered at a rate of not less than 700 gpm for any building required to have a total quantity of water less than 20,000 gallons and a rate of not less than 950 gpm for a building requiring a quantity not less than 40,000 gallons (Farrell, 1996).

- Fire Underwriters Survey (FSU): Similar to the ISO in the United States, the FSU has developed a "Guide to Recommended Practices" and is responsible for performing the review and evaluation of the adequacy of fire protection for municipalities and rating them accordingly. The FSU requirement for fire flow is based on building characteristics including the type of construction, floor area, number of stories, nature of occupancy, automatic sprinkler protection, and exposure risk. The minimum recommended fire flow is 500 gpm, and the maximum recommended fire flow is 12,000 gpm (O'Farrell 1996).

[Sidebar: with a few exceptions, this is the same criteria used by ISO in the USA.]

- Water supply standards: The water supply standards for the City of Edmonton, Alberta, Canada require a separate fire flow of 8 gpm to be imposed at a node adjacent to all high-value properties, e.g., schools and shopping centers. A separate fire flow of 1,600 gpm also is imposed at a node of the weakest link or farthest from the source of supply within the network for a single-family residential development. The residual pressure in all cases for any node within the network should not be less than 20 psi at ground level (Farrell, 1996).

Europe

There are no fire protection standards that apply to all European countries; therefore, each country must develop or adopt its own fire flow requirements.

Germany

The German standards for fire water requirements (DVGW 1978) are based on very detailed fire risk categories that assign the level of risk according to four classes of building use and three classes of fire spread risk (i.e., low, moderate, and high). German standards for building use categories are presented in Table 7-2.

Table 7-2: German Standards by Building Use Categories

Category	Description	Number of Stories	Floor Area Ratio*	Volume-Area Ratio
1	Small Building	_2	_0.4	—
2	Residential, light industrial, mixed residential, business mixed	_3	_0.3-0	—
3	Central Business District, Light Industrial	3	0.7-1.2	—
4	Central Business District, Heavy Industrial	1	1.0-2.4	_9

Source: Adapted from DVGW 1978.

* Floor Area Ratio = Total area of building/Footprint area.

Volume-Area Ratio = Total building volume/Footprint area.

The combined risk categories result in a fire flow requirements ranging from 200 gpm for the lowest risk area to 850 gpm for the highest risk area. The German standards require that all fire risk flows should be maintained for at least 2 hours with the residual fire pressure (the minimum pressure required during fire flow conditions) of 20 psi for all categories of risk.

United Kingdom

The flow rates that the UK Fire Services ideally require to fight fires are based on the national document on the provision of water for firefighting (Water UK and LGA, 1998.) Similar to German standards, fire flow requirements are determined based on the categories of premises to be protected. Table 7-3 shows UK fire flow requirements ranging from 120 gpm for the lowest risk areas to 1,200 gpm for the highest risk areas.

The UK guidance document does not stipulate the period for which the required fire flow for a specific risk category should be maintained, or the residual fire pressure.

Table 7-3 Categories of Premises and Fire Flows as Required in the UK

Category	Description	Fire Flow in gpm
1	Housing:	
	• Not more than two floor	Minimum of 250
	• More than two floors	300 to 550
2	Transportation (lorry/coach parks multistory car parks, service stations)	396
3	Industry:	
	• Up to 1 hector	300
	• 1 to 2 hectors	550
	• 2 to 3 hectors	800
	• Over 3 hectors	1,200
4	Shopping, office, recreation and tourism	300 to 1,200
5	Education, health and community facilities	
	• Village hall	250
	• Primary schools and single story Health centers	300
	• Secondary schools, college, and large Health and community facilities	550

Source: Adapted from Water UK and LGA ,1998.

Greece

The Technical Chamber of Greece (1996) stipulates that fire flows should be maintained for at least 30 minutes at a level that ranges from 200 gpm to 1,900 gpm depending on the risk category. The required residual pressure, which is 60 psi for all categories, indicates that the main rationale behind the residual fire pressure in this case may not be to act as a safety factor to prevent negative pressures developing in the water supply network (as in other countries), but to supply pressure directly to fire hoses.

France

The French standard for fire water requirement (Circulaire 1951, 1957, and 1967) require flows of 260 gpm supplied for 3 hours as a minimum. Similar to other European standards, fire flow requirements are determined based on the risk associated with premises to be protected, and in case of increased risk (e.g., factories, apartment buildings, etc.) the requirement will increase accordingly. The level of risk and required fire flows are determined based on assessment by the firefighting services. As a consequence of this approach, more stringent requirements can be imposed (Ruellan and Tiret, 1990). Such an example is given for the City of Paris where the requirements range from 530 gpm for a house with not more than two floors, to 2,640 gpm for industrial zones of importance to the general public. Although there were attempts in 1977 and 1978 to re-evaluate these standards, the standards described were still in use in France in 2002.

Spain

Similar to the French standards, the Spanish standards for fire water requirements (Real Decreed 1942/1993) require flows of 260 gpm to be supplied over 2 hours. The required minimum pressure is 13 psi.

Netherlands

In the Netherlands, water requirements historically are calculated based on now outdated KIWA Mededeling 50 (KIWA, 1977) according to defined risk areas a) the high-risk fire flows (required provision of 1,600 gpm to be supplied over 6 hours), or b) the low-risk fire flows (400 gpm to be supplied for 2 hours), which are the highest amount the European countries defined in the subject report. The required fire pressure specified in this document is 30 psi for all categories of risk. However, there are still 10 to 20 percent of the higher risk buildings that will require fire flows of 260 gpm. These higher risk standards are considered to be more of an exception than a rule. In the Netherlands, the fire department typically relies on the water utility for the first 15 to 30 minutes of a fire, after which time they use nonpotable water sources such as canals.

The Identification of Alternative Methods and Technologies for Structural Fire Suppression

Automatic Suppression Technologies

A host of structural fire suppression technologies were reviewed by both Committee and supporting staff that could have a positive impact on reducing water demand for structural fire suppression. The following is a synopsis of the technologies presented in the document under review for this manual. However, there has been significant advancement in several of these technologies along with better performance indicator since the AWWARF/KIWA report was published in 2002. Therefore much of the detailed information is left to be presented in Chapter 9–Viable Approach to Providing Adequate and Reliable Water Supplies for Fire Suppression and Chapter 10–The Latest Concepts on Water Supply Systems.

1. Automatic sprinklers.

Automatic fire sprinkler systems were the first and still are the foremost form of automated fire suppression system used throughout the world. With over 100 years of operating history and a 95 percent plus success record, these systems represent one of the most readily available means for effective fire suppression for a wide range of different occupancies. The success story with automatic sprinkler systems in confining, controlling, and extinguishing developing fires in structures has been a phenomenon.

When used in a closed-head sprinkler system, each individual fire sprinkler is, in a sense, its own fire detector and suppression device. A fire sprinkler is constructed of a mental frame with a threaded pipe inlet and a water distribution deflector; on the interior of the threaded inlet there is a discharge orifice through which the water discharges; a cap covers the discharge orifice. This cap is held in place by one of several types of heat responsive actuation mechanisms, (e.g., fusible link, chemical pellet, frangible bulb, or a special heat responsive element that opens and closes the orifice based on the heat intensity. The distinctions among the different types of actuation mechanisms is not significant relative to the sprinkler heads basic operation; only to the sensitivity of the operation. Each actuation mechanism has a predetermined (based on hazard to be protected and as defined by code) temperature set point. Once the mechanism has been heated to this set point, the mechanism will fuse, melt, rupture or other wise open the orifice allowing the water pressure in the piping system (in a wet pipe system) allowing water to discharge in a defined pattern over the fire.

Fire sprinklers have orifices ranging from 0.25-inch diameter up to 0.75-inch diameter, depending on the type of hazard or occupancy classification to be protected. Temperature actuation set points range from a low of 135 °F (57 °C) up to 650 °F (343 °C) Set points also are selected based on occupancy. Sprinkler water flow rate is a product of the square root of the water pressure at the sprinkler orifice multiplied by a K factor representing the sprinkler discharge characteristics. K factor ranges are from 1.3 up to 14.5, and even larger (for what are termed large drop-type sprinkler heads). The "standard" sprinkler as defined by NFPA 13 has a 1.2 inch (0.5) diameter orifice and a K factor of 5.6 and an actuation temperature of 155 °F (68 °C). At 15 psi, a "standard" open sprinkler head delivers just under 25 gpm.

Sprinkler systems are designed and arrange with special control valves that include wet-pipe systems, dry-pipe systems, preaction systems, and other special, less frequently used, systems. Wet-pipe sprinkler systems are by far the most common type of sprinkler system in service in commercial and residential properties today. Wet pipe, as the name implies, means that the piping distribution system to each sprinkler head is charged with water under pressure. It is important to note that in this type of closed head system, each sprinkler fuses individually in response to heat input from a fire. Only those sprinklers directly exposed to the heat of the fire will actuate to spray downward onto the developing fire. A common misconception is that when a sprinkler system actuates, all heads open simultaneously; this only happens with deluge systems, which are installed for very special hazardous situations. System actuation is alarmed through instrumentation at the system riser fed by a lead-in water main that detects water flow into the system and is wired to a local remote fire alarm panel.

Dry-pipe sprinkler systems operate similar to a wet-pipe system, except that the piping distribution system to feed the sprinkler heads is charged with air or an inert gas under a specified pressure. When an individual sprinkler head fuses, air or gas is discharged initially until the air pressure is depleted to a point where a differential type check valve located in the system riser will open and allow water to flow into the piping distribution system. Again, water flow is monitored by means of instrumentation to immediately sound an alarm. Dry-pipe systems typically are installed for the protection of areas where the sprinkler distribution piping is located in an unheated space such that if water were present, it could freeze and damage the piping, resulting in undesirable non-fire-related discharge of water. The design water flow requirements for a dry-pipe sprinkler system are basically the same as for wet-pipe systems except for the main control valve arrangement.

Preaction sprinkler systems are used in areas such as a rare book collection, where non-fire-related water discharge (e.g., due to sprinkler head or sprinkler piping) could be damaging. This system combines the design features of the dry-pipe sprinkler piping system and is charged with air or more likely. Today, with an inert gas with a separate fire detection systems in the protected space. Special pre-action control valve is located at the system riser and is tied to the fire detectors in the protected space. The preaction valve will not open to release water into the system until a fire alarm signal is received from the fire detectors. The signal will release the valve to admit water into the distribution system piping, but water discharge still will not occur until the heat of the fire has fused the actuation mechanism of an individual sprinkler head. Repeating, but this point is most important, water will only discharge from the sprinklers that have been heated to their actuation set point temperature. When the system is in a standby (non-fire situation) condition, the pressure of the inert gas (i.e., air) in the piping is monitored by a low-pressure switch. If a low air pressure condition occurs due to pipe damage or sprinkler head damage, a trouble alarm will sound so that repairs can be made prior to any water discharge.

2. Deluge sprinklers and water spray systems.

Deluge sprinklers and water spray systems operate in a manner similar to the preaction systems described above. The difference in these types of systems is that the actuation mechanisms in each sprinkler head has been removed so that all heads open on the system and water will discharge simultaneously. In both types of systems, fire detection devices are placed throughout the protected space and will actuate a normally closed deluge type valve at the system riser to flood the distribution piping and discharge water from all sprinklers or spray nozzles. Deluge sprinklers and water spray systems differ from one another only in that deluge sprinklers are standard type sprinkler head and are used for general floor area protection the same as other types of sprinkler systems. However, water spray nozzles are of a different design where the water pattern is directed at 30, 60, or 90 degrees from the nozzle orifice and are used for specialized applications such as the oil in transformers or to protect exposed structural steel. Both deluge sprinkler systems and water spray systems typically are limited to use in high-challenge, rapidly developing fire hazard applications. Deluge systems may be used to protect the storage of piled combustibles such as bailed cotton, cardboard containers, or containers on flammable and combustible liquids. These types of systems are for special hazard protection only. They are *inappropriate* for use in other types of occupancies due to water damage potential.

3. Foam-based sprinklers.

Wet-pipe, dry-pipe, preaction, deluge, and water spray type fire protection systems, when used for the protection of high challenging type fire hazards that include flammable liquid hydrocarbons, can be enhanced by the *addition* through injection of one of several types of fire fighting foam agents. The current basic foam classes include protein foam, fluoro-protein foams and Aqueous Film Forming Foam (AFFF) for flammable liquid fires and synthetic surfactant foams better know as Class A foams for ordinary combustibles and structural firefighting. These foams maybe used with fixed system applications and also are used commonly with manual hose streams and master stream applications in high hazard facilities.

The important concept here is that all of the firefighting foams are reported to extinguish fires in both ordinary combustibles faster and with less water than plan water alone; this reduces the water demand for firefighting. Foam in general tend to form a foam blanket that helps exclude oxygen and reduces vapor emissions from the fuel surface while allowing water droplets to settle out of the emulsion and provide cooling at the flame-fuel interface. While surfactant foaming agents are effective in controlling and extinguishing flammable liquid fires and there is increased applications of the newer Class A foams for more rapid control and extinguishment of developing structural fires in residential occupancies, this technology does not have a sufficient history of carefully documented successes to consider the reduction of needed fire flows at this time. Furthermore, the surfactant foams ultimately are carried to wastewater in the fire runoff. This may create an environmental concern regarding discharge of these "soaps" to wastewater streams in certain jurisdictions.

[Sidebar: See further details on Class A Foam in Chapter 9.]

4. Residential and fast-response sprinklers.

Within the past two decades, substantial decreases in the actuation time of sprinkler heads has been recognized. The design of sprinkler actuation mechanisms now incorporates a value known as the Response Time Index (RTI). RTI is a measure of thermal sensitivity, namely, the rate at which the sprinkler element can absorb heat from its surroundings, before actuating. RTI incorporates the concept of thermal lag in that while all actuation mechanisms are set to fuse at a given temperature, the mass of the actuating element introduces a lag time for heat flow into the element.

Significant research product development has led to two distinct types of state-of-the-art sprinklers, both having faster response times: the residential sprinkler and the quick-response sprinkler. Both residential and quick-response sprinklers incorporate lower mass fusible links or frangible bulbs that decrease RTI. The difference in the two types are in their discharge characteristics. Residential sprinkler systems have a modified water discharge pattern where more of the water discharged is contained in the upper portion of the horizontal discharge profile. This is necessary because residential sprinkler system design places a high reliance on the fire being suppressed with a single operating sprinkler head. The distribution pattern of the head must therefore protect draperies and higher elevation combustibles that are more typical to a residence than other types of occupancies. This discharge pattern also helps ensure that fire gases at the ceiling of an enclosure are cooled effectively, thereby reducing the likelihood of "flashover" and additional sprinklers actuating.

5. Water mist technology.

In the case of water mist fire suppression systems, it might be observed that something seventy years old is now a "new thing" again. Water mist technology was originally developed in Europe in the 1940s as application for marine engine and fuel space fire protection. In the 1950s a significant interest and research was direct at water mist as one of the agent most viable for replacing the Halon 1301 fire extinguishing agent. Federal regulations stopped the use of Halon 1301 by 1980 as a significant threat to Ozone depletion in the atmosphere. Beginning at the turn of the 21st century, significant interest and research has been directed at water mist as one of the agents most viable for replacing Halon in certain applications.

Water mist systems are defined by the different methods in which the mist is formed and the characteristics of the droplet size and distribution of the discharged mist from nozzles. Water mist droplets are formed through specially designed nozzles at pressures typically higher than those of standard sprinklers. The pressure is generated by high-pressure pumps or accumulators usually containing air or nitrogen, or even a combination of both. Water droplet size and discharge momentum are critical parameter in system design.

Additional Methods of Reducing or Augmenting Water Supplies for Fire Protection

The AWWA Research Foundation and KIWA Study on the impact of fire flows on municipal water distributions system also considered additional topics that include, 1) Surface Water Sources and Private Water Tanks, 2) Use of Non-potable Water Delivery Systems for Fire Protection, and 3) Continuing Sprinkler Technology. This study was completed in 2001 and published in 2002. Since that time there has been significant progress in these areas along with water mist technology that has been supported by the USFA in partnership program with the ISO, the National Institute of Science and Technology (NIST), the NPFA Committee on Suburban and Rural Water Supply, Standard 1142, and the Marriott Corporation. This update information is presented in Chapter 9 on Viable Approach to Reducing Needed Fire Flow or Providing Independent Water Supplies For Fire Protection.

Study Summary Statement of Importance

This referenced study provides the following conclusions based on the information through the year 2001.

Water-efficient fire suppression technologies suggest the use of less water than conventional standards. In particular, the universal application of automatic sprinkler systems provide the most proven method of reducing loss of life and property due to fire, while at the same time, providing faster response to the fire and requiring significantly less water than conventional fire fighting techniques. It is recommended that the universal application of automatic fire sprinklers be adopted by local jurisdictions.

Study Preparation By:

1. Jerry Snyder. Arun K. Deb, Frank M. Grabluta, and Sandra B. McCammon; Roy F. Weston, Inc., West Chester, PA, 19380.
2. Walter M. Grayman, Ph.D., Consulting Engineer, Cincinnati, OH.
3. Robert M. Clark, Ph.D., The United States Environmental Protection Agency, Cincinnati, OH.
4. Daniel, A Okun, Ph.D., University of North Carolina, Chapel Hill, NC.
5. Scott M. Tyler, Duke Engineering Services.
6. Dragan Savic, University of Exeter, Exeter, United Kingdom.

Project Advisory Committee

Note: Three committee sessions were held in the Philadelphia area and numerous phone contacts were made over a 2-year period. Some of the following individuals prepared separate documents that were used in writing the final study report.

- Mr. Isaac Pai, Long Beach Water Department, Long Beach, CA.
- Mr. Jonathan Clement, Black & Veatch, Boston, MA.
- Mr. William Kirkpatrick, East Bay Municipal Utility District, Oakland, CA.
- Mr. Jan Vreeburg, KIWA, the Netherlands.
- Mr. Jeffrey Swertfeger, Cincinnati Water Works (CWW), Cincinnati, OH.
- Mr. P.K. Tudor, Delaware County Water Company (DEL-CO), OH.
- Mr. Lewis Rossman, U.S. Environmental Agency, (USEPA), for modifications made to the EPANET software
- Mr. Scott Mitchell, American Fire Sprinkler Association (AFSA)
- Dr. Harry E. Hickey, Professor of Engineering, University of Maryland (Retired) representing the Insurance Services Office, Inc.
- Dr. Tom Walski, Haestad Methods, Waterbury, CT.

The authors would also like to thank the workshop participants for their input; these individuals include

- Dr. Yakir Hasit, Parsons Corporation
- Mr. Joan Hammel and Ms. Heidi Gorrill, Roy Weston, Inc.
- Mr. Michael Schmidt, ACE-INA Holdings

- Mr. Edward Straw, Senior Technical Engineer, Insurance Services Office, Inc.
- David Hauue and Mr. Hilosh Puchovsky, National Fire Protection Association (NFPA)
- Roland Huggins, ASFA, Mr. Timothy MacDonald, City of Cambridge Water Department
- Mr. Fern Marcuccio and Mr. Rob Cowan, Regional Municipality of Ottawa-Carleton
- Mr. David Hartman, CWW
- Mr. Ken Thompson, Irvine Ranch Water District
- Mr. Lou Chanin, United Water New Jersey
- Mr. Calvin Banning, Duke Engineering & Services
- Mr. Malcolm Brandt, Black & Veatch

CHAPTER 8: DUAL WATER SYSTEMS

Basic Concepts

The use of dual water distribution systems, one for potable water for consumer consumption and the other for nonpotable water is becoming a common practice. The main reason for this condition is the diminishing supplies of high-quality water resources, rapidly escalating costs for developing new water sources or for treating poor quality water, and, of no less importance, the increasing costs involved in discharging wastewater into the environment.

When faced with the task of developing additional water sources, community water utility managers and design engineers are increasingly evaluating the potential for dual water systems to serve their community's needs. Development of a dual water system maybe less costly and less wasteful than existing practices that use potable water for purposes that do no require the high quality of water prescribed by the EPA. Properly treated and distributed nonpotable water, as defined by the American Water Works Association (AWWA Manual M-24) can safely be used for irrigation, industrial applications, structural fire suppression, and a wide range of other nonpotable urban purposes to be identified in this Chapter. (1) The bottom line is that these practices conserve limited high-quality treated water for drinking, cooking and other uses requiring potable water that meets EPA quality standards.

Although several States have established regulations for the distribution and use of nonpotable water as defined in AWWA M-24, national standards and guidelines have not yet been established. The AWWA Distribution and Plant Operations Division Committee on Dual Water Distribution Systems, which prepared the stated manual, believes that more study is needed before national standards can be adopted. The following synoptic review of the AWWA concepts on dual water systems are of significant importance in providing needed fire flows without using domestic water. If this could become a reality, then water supplies for fire protection could be designed independent of consumer consumption requirements and provide avoidance for meeting consumption requirements along with needed fire flows on the same water system.

AWWA clearly states that it does not endorse the concepts and information on existing and hypothetical water systems presented in AWWA M-24. This is documentation for review, study, and comment. But the material provided needs to be considered carefully in relation to providing required water supplies by adopted building codes or needed fire flows by the insurance industry. It is also observed that a second approach to dual water supplies are not discussed in the AWWA M-24 Manual. To avoid a semantics problem, this second approach will be defined as a *Separate Water System* and will be discussed in Chapter 9.

Background Information

With increasing urban and industrial development, and with most high-quality sources of substantial water yield already developed, surpluses of high-quality water for the future are becoming less available, and in many areas of the country especially on the East and West Coasts they are nearly exhausted. Using *wastewater* treatment effluents for industrial, urban needs, or agricultural irrigation purposes is a very important way a nonpotable water supply can lessen the demand on potable water sources. An example is the introduction of wastewater treatment effluent for use in the steel industry in Baltimore in the 1950s. (2) Another example is the distribution of reclaimed wastewater to several industries in Colorado Springs in 1960 to relieve the demand on limited freshwater supplies.

Two primary factors have accelerated the development of dual distribution systems. First is the need for water supplies that do not necessarily need to be of potable quality for urban and industrial use. Second is new requirements for costly advanced wastewater treatment facilities, including nutrient and organic removal, to protect the receiving water supplies such as stream, rivers, lakes, and oceans.

Using reclaimed wastewater for nonpotable purposes may reduce water treatment costs when nutrient removal is not only unnecessary but wasteful. This is the case for urban and agricultural irrigation because nutrients already in the wastewater can replace those that would ordinarily need to be added. An example of this would be a golf course watered with reclaimed water. An extensive distribution system for urban irrigation, including the watering of parks, school and college campuses, median strips, and residential lawns, is an important element of reclaimed wastewater system used for other nonpotable purposes.

Although there are many dual distribution systems of varying degrees of complexity, the Federal government has not yet established guidelines for nonportable distribution systems. Some State governments have established standards and guidelines for reclaimed water quality and its nonpotable use. This Chapter present a general summary of current practices for dual water distribution systems with a special focus on the application of these systems to providing an alternative approach to providing required and needed fire flows for structural fire suppression operations. A limited application of this concept is found in the States of Arizona, California, and Florida.

Potential Applications for Dual Distribution Systems

With many successful dual water distribution system now in operation, running from systems that are just one or two major customers to systems that serve all properties in the community, water supply professionals are now considering dual distributions that address special water supply demands and water pollution control problems. These special problems including providing water for fire protection may necessitate new piping system configurations and the engineering of pipe sizes to meet fire flow requirements. This raises the issue of the need for more than one type of reclaimed water distribution system for a given community. The following situations in which a dual water distribution system would be worth evaluating are described under the following topics.

1. Limited water resources.

Competition of existing water resources or limitations on these resources often make acquisition of additional water resources politically and financial difficulty. For example, obtaining additional water resources might require inter-basin (raw water) transfer, which is less politically acceptable now than it was in earlier years of water resource development. Greater knowledge of groundwater hydrology has illuminated the danger of excessive withdrawls from aquifers, which can result in continually increasing costs of pumping, impaired water quality, and land subsidence. Limited resources create a situation in which water reclamation and a dual distribution system may become attractive.

2. Limited water supplies.

When demand is expected to exceed the yield from existing water supply facilities, and additional facilities need to be constructed, the gradual introduction of a nonpotable water might be appropriate. Water users, whether industrial, commercial, or municipal, may well be served with reclaimed wastewater in place of potable water at a cost substantially lower than the cost of developing new high-quality water sources of supply for potable purposes.

3. Polluted sources.

The 1982 United States Public Health Service Drinking Water Standards recommended that "water supply should be obtained from the most desirable source which is feasible..." (USPHS 1982). The EPA National Interim Primary Drinking Water Regulations adopted in 1995 state that "priority should be given to the selection of the purest source." Many communities have selected the highest quality available which varies widely by region of the country. Additional development may require use of a more polluted source which extrapolates the cost of water treatment to meet EPA regulations. In the past, engineers believed that they could render almost any polluted water safe for potable supply through adequate treatment; generally from coagulation, filtration, and disinfection. However, the "chemical revolution" in the United States since the 1950s resulted in thousands of synthetic organic chemicals that are not readily degraded in the environment. The number of contaminants regulated in potable water has grown from 4 in 1925 to a projected legislated number exceeding 150 by 2000. The treatment of water that will meet drinking water standards in the future will cost considerably more than it does in 2005. Polluted sources of water may therefore be less suitable for potable water supplies but well suited for one or more nonpotable purposes.

4. Rigorous wastewater treatment.

To restore and preserve the quality of North American water, treatment requirements for community wastewater have become increasingly stringent and costly. In many cases, nutrients may need to be removed. If the wastewater can be used in industry for urban irrigation, some advanced treatment, including nutrient removal, may not be necessary and may in fact be undesirable. The operation and maintenance of advanced physical and chemical wastewater treatments facilities are generally more costly than treatment for nonpotable reuse. If wastewater can be used productively, the savings in wastewater treatment can be passed on to users, making water reclamation and dual distribution systems more economically attractive.

Sources of Nonpotable Water

The most commonly used source of nonpotable water is water reclaimed from local wastewater plants, which is generated near customers that can use it. The technology needed to assure the required quality of water is well established. Treatment of secondary effluent is accomplished by coagulation, conventional filtration, and disinfection. Other sources of nonpotable water include brackish or mineralized surface waters and groundwater, including seawater, certain industrial wastewaters and ground waters, and irrigation return flows. The focus of this Chapter is on reclaimed water from sewage treatment plants. Chapter 10 discusses completely *separate* independent water supplies for fire protection that include streams, rivers, both small and large lakes, and ocean water.

Potential Uses of Nonpotable Water

A significant objective of this Water Supply Concepts Manual is to examine ways to reduce required or needed fire flows **or** examine viable approaches to providing water for structural fire suppression without using water treated to EPA specifications for water treatment. This is to reduce the demand for high-quality potable water and to reduce the operating costs for water treatment facilities and the distribution pipe sizes needed to supply domestic water at peak demand **and** needed fire flows simultaneously.

It is very important to understand that water for fire protection is not the only requirement that drives the size and cost of water treatment plants and distribution system design. Typical discussion of water treatment is confined to the need of treated water to meet health standards, for residential use, hospitals, nursing homes, restaurants, and basically where water is involved in the human consumption process. Forgotten are the myriad uses of water by the Federal government, State government, county government, local community government, commerce and industry, and homeowners that may not require a high quality of water.

Potential Uses for Nonpotable Water

A dual water system that reuses treated sanitation water under less stringent criteria than potable drinking water has many uses that can significantly reduce the need for potable water from traditional municipal water delivery system. Some examples include the following:

1) Public uses:
   a) Park irrigation and water amenities, such as ornamental fountains.
   b) School or campus landscaping and playground irrigation, median strips and roadway right-of way landscaping irrigation.
   c) Recreational facilities, such as gold-course irrigation, tennis court wetting, washing, and ski-slope artificial snow.
   d) Cemeteries and nurseries.
   e) Public facility toilet and urinal flushing.
2) Public and Private Sector Fire Protection:
   a) Separate water system to provide fire hydrant water that is independent of the municipal water supply system to provide better adequacy and reliability in terms of designed flow rates at specified fire risks and pressure to supply fire department pumpers at substantially lower operating costs than is possible with combined systems.
   b) Providing piped water systems to cope with the Urban/Wildland interface problem in many States. Provide water supplies for fixed fire protection systems including foam system, water spray systems, standpipe system. Note: Probably should not be used for commercial or domestic automatic sprinkler systems.
   c) *** Key Consideration: A dual water system would improve the reliability of a water system for homeland security. If the municipal water system was contaminated or disrupted, there would be a secondary supply that could be used with portable purification units to supply domestic water as well as providing water for emergency services.
3) Industrial and Commercial Uses:

    a) Cooling-tower make-up water.

    b) Boiler feed water make-up water.

    c) Stack gas scrubbers.

    d) Process water.

    e) Crop irrigation.

    f) Construction, including concrete manufacturer, soil compaction, and dust control.

    g) Toilet and urinal flushing.

    h) Street cleaning and car washing.

4) Residential uses:

    a) Lawn watering.

    b) Toilet flushing.

All of these applications are being practiced. For example, toilet flushing with reclaimed water has been practiced in Grand Canyon Village, Arizona, since 1926. A relatively recent development in Singapore is the use of reclaimed water, originally intended for industrial process water for toilet flushing in a highrise housing estate for some 25,000 people since 1991 (3), reclaimed water has been used for toilet flushing in several highrise buildings in the Irvin Branch, CA, Water District. A growing practice in the United States is the reclamation of wastewater originating in shopping malls and office complexes for toilet flushing in the same complexes. This practice reduces the volume of wastewater discharge.

Other uses of nonpotable water include enhancing wetland and ground water recharging.

Nonpotable Water Reuse Legislation

The rapid growth in the number of nonpotable water reuse project indicates that the potential for nonpotable water reuse *needs to be* evaluated whenever communities need additional water supplies for domestic consumption and/or fire protection. In three principal States, California, Florida, and Arizona, nonpotable water reuse is stimulated through legislation. It is important to understand the current state of this legislation which is in the foundation stage for moving the application of Dual Water Systems using reclaimed water forward to meet Federal and State requirements to protect public health, and with special attention to using this type of water supply for fire protection purposes.

The following is a summary of State reuse regulations based on a survey of all States conducted by the EPA to develop guidelines for reuse of water. Appropriate regulatory agencies in all 50 States were contacted by mail in September 1990 by EPA and asked to provide information concerning their current regulations governing water reuse. After the followup contacts, all 50 States responded to the request for information. All of the information presented in below is considered current as of March 1992. The literature from the AWWA does not indicate an update on this information.

As of the stated study date above, it was determined that most States do not have regulations that cover all the potential uses of reclaimed water. The notable exceptions are Arizona, California, Florida, Texas, Oregon, Colorado, Nevada, and Hawaii, which have extensive regulations prescribing requirements according to the various uses of the reclaimed water. Other States may have regulations or guidelines that focus on land treatment of wastewater effluent. These guidelines emphasize additional treatment of effluent disposal rather than beneficial reuse, even though the effluent may be used for irrigation or agricultural sites, golf courses, and public-access lands.

Based on the survey, current regulations generally are divided into the following reuse categories:

♦ Unrestricted urban use for irrigation of areas in which public access is not restricted. This includes such areas as parks, playground, schoolyards, and residences; and uses such as toilet flushing, air conditioning, building construction, ornamental fountains, aesthetic impoundments, such as reflection pools, and urban and rural fire protection.

♦ Restricted urban reuse: irrigation of areas which public access cannot be controlled, such as golf courses, cemeteries, and highway medians.

♦ Agricultural reuse of food crops: irrigation of food crops intended for human consumption. This category is often further classified as to whether the food crop is to be processed or consumed raw.

♦ Agricultural reuse on nonfood crops: irrigation of food fodder, fiber and seed crops, pasture lands, commercial nurseries, and sod farms.

♦ Unrestricted recreational reuse: an impoundment of reclaimed water for which no limitations are imposed on body-contact, water recreation activities.

♦ Restricted recreational reuse: an impoundment of reclaimed water for which recreation is limited to fishing, boating, and other noncontact recreational activities.

♦ Environmental reuse: reclaimed water used to create artificial wetlands, enhance natural wetlands, and to sustain stream flows.

♦ Industrial reuse: reclaimed water used in industrial facilities primarily for cooling-system makeup, boiler feed water, process water, and general wash down.

Reclaimed Water Quality and Treatment Requirements

Requirements for water quality and treatment of water reuse receive the most attention in State regulations. States with water reuse regulations or guidelines have set standards for reclaimed water quality, specified minimum treatment requirements, or both. Generally, when unrestricted public exposure is likely in the reuse application, wastewater must be treated to the highest degree possible before its application. Where exposure is not likely, however, a lower level of treatment usually is accepted. The number of States with regulations or guidelines on this subject is presented in **Table 8-1**.

Water quality limits are commonly imposed on biochemical oxygen demand (BOD), total suspended solids (TSS), total or fecal coliform counts, and turbidity. Coliform counts and turbidity are generally used as indicators to determine the effectiveness of disinfection. A limit on turbidity is usually not specified or required unless there is a need to achieve a high level of disinfection.

Table 8-1: Number of States with Regulations or Guidelines for Each Type of Reuse Application

Types of Reuse	Number of States
Unrestricted Urban Irrigation	22
Toilet Flushing	3
Structural Fire Protection	2
Building Construction	4
Landscape Impoundment	7
Street Cleaning	1
Restricted Urban	27
Agricultural (Food Crops)	19
Agricultural Non-food Crops	35
Unrestricted Recreational	5
Restricted Recreational	7
Environmental Wetlands	3
Industrial	6

Reclaimed Water Monitoring Requirements

Reclaimed water monitoring requirements, as well as the type of reuse, vary greatly from State to State. For unrestricted urban reuse, Arizona requires sampling for fecal coliform on a daily basis, but for agricultural reuse of on nonfood crops, such sampling is required only once a month. Arizona also requires turbidity to be monitored on a continuous basis when a limited turbidity is specified.

California, Florida, and Washington also require continuous online monitoring of turbidity. Oregon, on the other hand required that turbidity be monitored hourly and the sampling for total coliform be conducted either once a day or once a week, depending on the type of reuse application.

California requires total coliform samples to be taken on a daily basis and turbidity to be monitored on a continuous basis. For unrestricted urban reuse, as well as agricultural reuse on food crops, Florida requires the continuous online monitoring of turbidity and chlorine residual. Even though no limits on turbidity are specified in Florida, continuous monitoring serves as an online surrogate for suspended solids. In addition, Florida requires that the TSS limit must be achieved prior to disinfection and the fecal coliform samples be take on a daily basis. Other States determine monitoring requirement on a case-by-case basis, depending on the type of reuse.

Treatment Facility Reliability

To achieve reliability in reclamation facilities, some States have adopted regulations or guidelines to ensure that reclaimed water of required quality will be delivered. Methods of ensuring reliability consist of alarms that warn of power failure of essential unit processes, automatic standby power sources, emergency storage, and the provision that each treatment process be equipped with multiple units or a backup unit.

California Title 22 Regulations provide design and operational considerations covering alarms, power supplies, emergency storage and disposal, treatment processes, and chemical supply, storage, and feed facilities. For treatment processes, a variety of reliability features are acceptable in California.

Groundwater Monitoring

Groundwater monitoring programs associated with irrigation of reclaimed water are required by Arkansas. Delaware, Florida, Georgia, Hawaii, Illinois, Maryland, Missouri, Montana, New Jersey, South Carolina, Tennessee, Washington, West Virginia, and Wisconsin. In general, these groundwater monitoring programs require that one well be placed up-gradient of the reuse site to assess background and incoming groundwater conditions with the aquifer in question. In addition, two wells are to be placed down-gradient of the reuse site. Some States also require that a well be placed within each reuse site. Florida generally requires a minimum of three monitoring wells at each reuse site, although for systems that involve unrestricted urban reuse, they require a plan to monitor typical areas. South Carolina's guidelines suggest that a minimum of nine wells be placed on 18-hole golf courses that irrigate with reclaimed water. Sampling parameters and the frequency of sampling are generally considered on a case-by-case basis.

Treatment for Reclaimed Water

The latest documented information on the general state of water regulations and basic support information on this topic is presented above. Now it is important to present some of the specific aspects of water for nonpotable use. The focus of the following information applies specifically to reclaimed water but also references not potable treatment for other uses as specified.

The major water-quality parameters include biodegradable organics, suspended solids, plant nutrients, dissolved solids, and pathogens. The required treatment of wastewater depends on the highest use of reclaimed water in the system. Usually reclaimed water is treated to meet unrestricted urban reuse, which requires the highest-quality reclaimed water.

For example, water for body-contact recreational use could require a high degree of removal of all constituents except dissolved solids. Orchard irrigation via surface furrows could require relatively lower degrees of all removals and probably no intentional removal of nutrients or dissolved solids if wastewater is the source of supply. Certain industrial water demands, such as water for cooling towers, could require a very high degree of removal of dissolved solids, including conversion of ammonia and phosphorus, and silica removal. The following information gives an overview to the treatment process:

Basic Treatment Process

1) **Biodegradable organics:** Biodegradable organics are often measured as biochemical oxygen demand (BOD), chemical demand (COD), and total organic carbon (TOC). Starting with wastewater, gravity sedimentation removes 30 to 40 percent BOD, producing water acceptable for flood irrigation. Biological processes remove 85 to 98 percent BOD, producing water acceptable for landscape irrigation, Chemical coagulation and filtration remove greater than 99 percent of BOD, producing recreation water. Activate carbon removes virtually all BOD, producing water that may be suitable for groundwater injection, or the replenishment of groundwater. This is a very important concept as groundwaters in most areas are receding; in fast-growing residential areas the recession process is alarming.

2) **Suspended solids:** Suspended solids often are measured as total or volatile suspended solids in all but very low concentrations (i.e., less than 5 mg/L), otherwise turbidity is used as an indication of solids concentration, and the percentage removal from wastewater is approximately the same as with BOD.

3) **Plant nutrients:** Plant nutrient is generally not beneficial for irrigation uses, where the nutrients effectively replace commercial fertilizer. Removal might be beneficial for certain industrial process where inhibition of biological growths or corrosion control is important. Nutrient-removal processes are usually specific to a given nutrient, such as P or N, or to the specific form of the nutrient, Examples would be nitrification to convert ammonia-N to nitrate-N and identification to convert nitrate-N to nitrogen gas.

4) **Dissolved solids:** Dissolved solids usually include inorganic constituents, major anions, and cautions. They are often measured indirectly as electrical conductivity, directly as dissolved solids by evaporation, or partially as specific ions. Removal of dissolved solids is beneficial to some extent for irrigation reuse and to a greater extent for various industrial uses. Dissolved-solids-removal processes vary greatly, both by method and effectiveness. Some remove dissolved solids in general, as with reverse osmosis, while others remove only one constituent, as with selective ion exchange.

5) **Pathogens:** The presence of bacterial and viral pathogens and sometimes parasites, is often measured indirectly by an indicator organism, such as coliform bacteria. The removal of pathogens is necessary for all reuses. The degree of removal required depends on the specific reuse, including the extent of direct and indirect exposure of the general public to the reclaimed-water supply. Disinfection is usually accomplished with chlorine, although other methods, such as ozoneation and ultraviolet-light radiation, are being used.

Reclamation Plants

Reclaimed water can enter the nonpotable water distribution system from a wastewater treatment plant, which provides secondary treatment as well as filtration and disinfection. Secondary treatment consists of screening and grit removal sedimentation, and low-rate treatment processes, such as stabilization ponds and aerated lagoons; or high-rate processes such as activated sludge, trickling filters, or rotating biological contractors. Experience with such plants demonstrates that a high quality secondary effluent, treated with small doses of coagulant, polymer, or both, direct sand filtration, or chlorine disinfection can meet unrestricted urban reuse requirements (Okum, 1991).

One advantage of using reclamation plants is that they can be located in the area where the reclaimed water is needed. Excess flow, excess reclaimed water, and the sludge can be returned to the trunk sewer to be handled at the central wastewater plant.

Design aspects of reclamation plants include standby power supplies, standby or replacement equipment, treatment-process redundancy, emergency storage or replacement equipment, treatment-process redundancy, emergency storage and disposal provision (including alarm systems), monitoring equipment, and automatic actuators. Operational aspects include the development, documentation, and exercise of sound routine and contingency operation procedures, and a comprehensive preventative maintenance program. All of these considerations help to ensure that only water of acceptable quality is delivered to the reclaimed water distribution system.

Storage

Matching the nonpotable water system supply with demand is important, particularly when wastewater is the course of supply for nonpotable water. Wastewater flow generally is constant throughout the year, while nonpotable demand (typically irrigation water and possibly water for fire protection) has irregular and instantaneous flow demand. The variance between supply of reclaimed water and demand can be compensated in several ways including the following:

◆ store enough nonpotable water to balance supply and demand;

◆ supplement nonpotable water with another source to satisfy peak demand;

◆ dispose of excess nonpotable water; and

◆ a combination of the above.

The reliability required of a nonpotable water system is usually not as great as that of a potable water system. This is primary true where nonpotable water is used for irrigation purposes, street cleaning, and other intermittent uses. However, if nonpotable water is to be used for toilet facilities, fire protection, and industrial process water that reliability has to design for 24-7 continuous use at specified design requirements.

Where fire flows may be provided by the nonpotable water source system, storage should be sized to meet required or needed fire flows for the representative fire risk locations. A second design possibility is to have the potable water system meet a minimum fire flow demand of 500 gpm at 20 psi throughout **ALL** built-upon areas of the community, and use treated reclaimed water to meet the difference between 500 gpm and needed fire flows (i.e., the difference between 500 gpm standard availability on the domestic system and a NFF of 1,500 gpm would be 1,000 gpm to be supplied by the reclaimed water supply). *This is an extremely important concept!*

One underlying focus of this manual is to explore possible methods of reducing required water supplies by building code and or needed fire flow developed by the ISO. Instead of considering the reduction of required or needed fire flows, it is equally important to consider alternatives for meeting the structural fire suppression water supply needs. Reclaimed water is one important way to augment existing water supplies to provide water for fire protection without enhancing a water system based on fire flows. In facts with fire flow reductions, a given water system may be more adequate to meet consumer consumption demands without needing improvements to the potable treatment plant or the distribution system to meet both consumer consumption and fire flow simultaneously.

This would require new diligence in fireground operations. Fire department pumpers operating on the domestic fire hydrant system *cannot* have supply lines also coming from a fire hydrant on a reclaimed or essentially a nonpotable water system. The reclaimed water supply **CANNOT** contaminate the treated water supply. Where the types of systems co-exist, there needs to be development and adherence to standard operating procedures. Important considerations on this matter are presented under Reclaimed Piping Systems below.

Features of Reclaimed Water Distribution Systems

The general features of Dual Water Systems as defined in this Manual which coincides with AWWA-M–24 basically considers water treated piping systems for the irrigation, lawn watering car washing, street cleaning, and some selective industrial uses with more recent attention to providing water in commercial buildings for sanitary requirements. The distribution systems are hydraulically designed to the end user requirements which typically are low flow because irrigation is done on alternative days and the same applies to watering lawns, shrubs, greenways, etc. Constant users are car washes, and some industrial applications. Therefore, pipes along streets may be as small as 2 inches, and many systems have pipe no larger than 4 inches.

This presents a problem when reclaimed water is to be considered for fire protection. The only documented used of reclaimed water for fire protection is in Florida, California, and Arizona. It should be noted that all three States have essentially nonfreezing weather where this application is in evidence. This allows for the use of wet-barrel hydrants and the lack of requirements to bury pipelines below the frostline. The only real requirement is to provide sufficient cover for loading under roads and driveways. Since no overall standards or engineering documentation are in place concerning these issues, an actually reclaimed-water fire hydrant system used in southern Florida is presented as one possible approach to augmenting existing treated water supplies to provide water for the wildland/urban interface problems and to provide "make-up" water to meet needed fire flow in high value districts and even residential districts when there is a water shortage problem.

The waste water treatment facility is located just west of the Intercoastal Waterway. One pipeline parallels U.S. Route 1, one follows a major boulevard, and one tracks along the east side of I-95. Wet-barrel fire hydrants are located approximately every 1,000 feet along the primary line located 15 feet from the edge of the roadways. Hydrants along the I-95 corridor are located on the outboard side of the northbound lanes, 5 feet away from guard rails and 1 mile apart. The intent of the hydrant distribution along the Interstate highway is to provide water for natural cover fuel fires and other fire emergencies on the highway plus as source for refilling water tankers.

Hydrants along Route 1 and the boulevard are used to protect both commercial and residential developments along with protecting boating marinas along the coastal waterway. Smaller diameter pipes, ranging from 4 inch down to 2 inch, cross-connect along major streets that run east and west in order to provide water for lawn irrigation, car washes, street cleaning, and dry vegetation spraying. To avoid any possible confusions with fire hydrants on the treated potable water system, the fire hydrants are painted purple and have posted signs indicating that the water is "Treated Waste Water." The greatest fire protection use for this system has been in the control and extinguishment of brush fires in highly populated areas.

While such a system as described above is practical and rather economical to install in non-freezing climates such installation in northern climates would require frostproof fire hydrants, much deeper pipe trenches, and valving. However, the concept still has much merit as the demand for potable water increases. Care must be exercised in the design to make sure that the desired water for fire protection use is available when there is peak demands on the system for irrigation or the watering of lawns.

The Future Role of Dual Water Systems

Dual water systems, must of necessity, become larger role players in providing water supplies to populated areas to meet a significant portion of the demand for water. This Chapter has identified the many uses of reclaimed water through proper treatment. One of those area to be carefully evaluated in the future is the role dual water systems can play in meeting required and needed fire flow demands which could take away the necessity for expensive new potable water treatment plants and water main enlargements to meet both domestic water supply needs and fire flow needs.

In new communities or in the expansion of communities, consideration also can be given to shifting the water supply for fire protection to the reused water supply. This would require designing the pipe distribution system to provide needed fire flows for representative risk location at 20 psi residual pressure while maintaining the other demands such as irrigation that may be in operation at the same time. The big payoff is that this would leave the domestic water system with the challenge of just meeting consumer consumption requirements *without* meeting needed fire flows. Re-channeling the supply of water for fire protection may be a more reasonable and responsible approach to the water supply crisis than focusing on reducing the requirements and needed fire flows during extreme dry weather and to provide fire breaks.

References:

1. AWWA M–24. *Dual Water Systems*, 1994 Ed. American Water Works Association, 6666 West Quincy Ave., Denver, CO 80235.

2. Okuun, D. A. "Planning for Water Reuse." *Journal AWWA*, 65(10): 617.

3. Okuun, D. A. "Reclaimed Water: An Urban Resource." *Water Science Technology*, 24(9): 353, 1991.

CHAPTER 9: WATER SUPPLY AND EFFECTIVE FIRE PROTECTION

Overview

Most communities in the United States consider that *growth* is a good thing. Development brings prosperity to cities, communities, suburbs, and the rural environment. And many parts of America are experiencing *extraordinary* residential and commercial growth.

But, growth can also strain community resources that include but are not limited to fire alarm, and emergency communications, fire department response to structural fires and other emergencies, and, most importantly the community water system. New construction, increasing population, and expanding economic activity places demands on a wide variety of municipal services including the above three elements of public sector fire protection. In many places, the need to enlarge public services, especially those functions related to fire protection, comes at the time when other priorities are competing for the same budget dollars.

The results can be devastating. If the fire department, emergency communication facilities, and *water supply* cannot keep up with the demand for services, a growing community faces the risk of property loss and personal injuries. The purpose of this Chapter is to examine the importance of water supply in the equation of providing effective fire protection at the local government level. This includes looking at water for structural fire suppression provided by fire departments and installed fire protection provided by automatic sprinkler systems and the new water application technology–*water mist systems*. Performance considerations and economic considerations are taken into account in examining the following topic headings.

The Insurance Services Office, Inc.

Property/Casualty insurance companies have long supported the efforts of individual communities to maintain and improve there fire protection services. ISO's Public Protection Classification (PPC) program helps insurance companies measure and evaluate the effectiveness of fire-mitigation services throughout the country. (1) The PPC program also offers economic benefits in the form of lower insurance premiums for communities that invest in the firefighting services that are evaluated in determining PPCs on a scale of 1 to 10. The Community Classification is determined by evaluating the 1) public fire alarm and communication capability for receiving and handling fire and emergency calls, 2) public fire department capability to confine, control, and extinguish developing fires in structures, and
3) adequacy and reliability of the municipal water supply to deliver needed fire flows at representative fire risks throughout the community.

A given community is surveyed by ISO Field Representatives using the ISO *Fire Suppression Rating Schedule*. The weighted credit is assigned as follows: (1)

　　　　1) Receiving and Handling Fire Alarms........ 10 percent

　　　　2) Fire Department...................................... 50 percent

　　　　3) Water Supply40 percent........................... <u>40 percent</u>

　　　　　　　　　　　　　Total　　100 percent

ISO is an independent organization the serves over 1,300 property and casualty underwriters in the United States. In addition, ISO directly serves fire departments, insurance regulatory agencies, and others about fire and casualty risk levels. ISO's expert field staff visits communities around the country to collect information about the items evaluated under the *Grading Schedule* items as titled above. For each of more than 46,000 fire districts in the United States, ISO analyzes that information and assigns a PPC on a scale of 1 to 10. Class 1 represents exemplary fire protection, and Class 10 indicates that the area's fire suppression program does not meet ISO's minimum criteria; this is an "unprotected" classification for insurance rating purposes. Of special *note* is Class 9 properties. This class includes all structures that are within 5 travel miles of a fire station but are located more than 1,000 feet from a recognized fire hydrant that flows a minimum of 250 gpm. According to a recent survey by the ISO 34.8 percent of the 46,000 fire districts in the United States do not have a recognized water supply for fixed protection; that is in excess of 1/3 of the land surface of the United States. (1)

The Ability of Fire Departments to Provide Effective Fire Protection

Even in well-protected areas, growth often threatens the ability of fire department to keep up with the demand for service. (1)

Many fire chiefs believe that residential and commercial growth is threatening the ability of fire departments to protect the communities they serve. Without careful attention, long-term planning, and adequate funding, communities risk deterioration in their protection, especially in these fast-growing areas. The threat is real, even in communities where fire protection is adequate today.

Those are the most important findings from a survey of fire chiefs, water superintendents and other municipal officials conducted for ISO by the independent polling firm Opinion Research Corporation (ORC) of Princeton, New Jersey, and published in 2003. (1) ORC interviewed 500 top-level municipal officials from randomly sampled communities around the country. The survey examined current conditions as well as development likely to occur over the next 3 years. The margin of error for survey results is plus or minus 2 percentage points.

About 86 percent of the fire officials reported that their communities had experienced residential or commercial growth over the past 3 years. In those communities, almost three-quarters (73 percent) of the respondents said that the growth is straining their departments' ability to protect the communities. Asked where growth had occurred, 71 percent of the chiefs said that it occurred in areas where they needed or still need improvement to water mains, fire hydrant distribution, or hauled water services. And 50 percent said that growth had occurred in areas where they needed or still need additional fire stations.

[Sidebar: The term "hauled water" has the same reference as alternative water supplies as used in this Manual (Chapter 6).]

Fire Insurance Rates

Property and casualty underwriters use two basic factors to establish fire insurance rate premiums:

1) Fire losses or loss costs.
2) Fire protection jurisdiction PPC.

Only loss costs are used for cities of over 500,000 population; fire losses are evaluated by counties and States. (1) It should be obvious that the adequacy and reliability of water supplies has a very significant impact on both life loss and property loss reduction at the time of a fire and the PPC. If water supply does not meet demand, it could have an adverse effect on insurance premiums for both commercial property owners and homeowners—more for commercial properties and than homeowners. Conversely, the ability to meet needed fire flows in both hydrated nonhydrated areas most likely will result in premium reductions if all factors of the fire protection system keep pace with urban and suburban growth.

An Overview of Water Supply Under the ISO Fire Suppression Rating Schedule

It appears important to have a relative understanding of the items that are reviewed by the ISO Field Representatives in conducting municipal water supply evaluations. Copies of the most recent survey of a water system can be obtained from the Regional Office of ISO found in the Reference Section to this Chapter. The following information provides the highlights of each item considered, by title. (2)

- **Part of the City (Community) Unprotected:** If any part of the city is not within 1,000 feet of a recognized water system, the unprotected area may receive a Class 9 rating. A recognized water system must delivery 250 gpm for a duration of 2 hours.

- **Maximum Daily Consumption Rate (MDC):** The maximum daily consumption rate is the average rate of consumption on the maximum day. The maximum day is the 24-hour period during which the highest consumption total is recorded in the last 3-year period. High consumption that will not occur again due to changes in the water system, or water loss caused by unusual conditions (i.e., a water main break lasting less than 1 hour) will not be considered.

- **Minimum Pressure:** A water system is reviewed at a residual pressure of 20 psi.

- **Fire Flow and Duration:** The fire flow duration should be 2 hours for needed fire flows up to 2,500 gpm, and 3 hours for needed fire flows of 2,000 and 3,500 gpm as determined under Section I of the *Grading Schedule*. Requirements under Section II of the *Grading Schedule* call for a 4-hour duration for fire flows ranging from 4,000 gpm to 12,000 gpm.

- **Service Level:** A service level is a part of the city distribution system that is served by one or more sources of supply but that is separated from the remaining distribution system by closed valves, check valves, or pressured regulating equipment, or is not connected.

- **Review of Supply System:** The ability of the water supply system to deliver needed fire flow at representative locations throughout the city is reviewed under the following bullet items. For each representative location, the supply works, mains, and hydrant distribution are reviewed separately.

- **Supply Works:** The absolute minimum supply available from water sources under extreme dry conditions should not be taken as the measure of the normal ability of the source supply. The normal sustained flow of supplies should be used as the normal capacity of the water source. If the water supply is regularly reduced for a period exceeding 1 month, the available supply is to be prorated by the time it is available. This evaluation covers:
  - Minimum storage
  - Pumps
  - Filters
  - Emergency supplies
  - Alternative water supplies (i.e., See Chapter 6)

- **Water Main Capacity:** The normal ability of the distribution system to deliver needed fire flows at representative risk locations throughout the community, as determined by the ISO are to be evaluated. The results of flow tests at each representative test location will indicate the ability of the water mains to carry water to that location. The AWWA/ISO methods of flow testing are covered in Manual II on *Water Supply Evaluation Concepts*.

- **Fire Hydrant Distribution:** Only fire hydrants located within 1,000 feet of a representative risk location measured as fire hose can be laid by responding fire apparatus is credited to satisfy needed fire flow requirements. Maximum credit for each fire hydrant within 1,000 feet of and identified fire risk is as follows: 1) Credit up to 1,000 gpm for each hydrant within 300 feet of the location, 2) 670 gpm from hydrants within 301 to 600 feet of the location, and 3) 250 gpm from hydrants within 601 to 1,000 feet of the risk location. Furthermore, the maximum credit for each fire hydrant maybe limited by the lack of a pumper connection and the number of 2-1/2-inch house outlets and a minimum water delivery capability of 250 gpm.

- **Fire Hydrants–Size, Type and Installation:** Point awards for installed and connected fire hydrants to water delivery piping of 4 inch or larger are prorated according to the following chart:

Installation Condition	Points
with 6" or larger branch, a pumper outlet and 2 to 2-1/2" outlets	100
with 6" or larger branch, no pumper outlet and 2 to 2-1/2" or more outlets	75
with 1 to 2-1/2" outlet only	25
with less than a 6-inch branch	25
Flush type	25
Cistern or suction point	25

- **Inspection and Condition of Fire Hydrants:** The inspection and condition of fire hydrants should be in accordance with AWWA Manual M-17–*Installation, Maintenance, and Field Testing of Fire Hydrants*. (3) To receive full credit under this item, fire hydrants need to be inspected every 6 months. Reduced prorated credit is given for less frequent inspection and flow testing up to 5 years. After 5 years a hydrant is considered "unreliable" and no credit is given for the fire hydrant in protecting a specific fire risk. (3)

Insurance Company Assistance on Evaluating Water Supplies

The insurance industry can assist communities in the following ways to assess the current water supply demand requirements:

⬥ ISO can provide an identified community with a list of the needed fire flows for representative fire risk, along with available fire flows at the time of the last community PPC Survey.

[Sidebar: See appropriate contact information through the Regional ISO Offices identified under the references.]

⬥ ISO can provide an identified community with a copy of the last PPC Report that covers all elements of the water supply system.

⬥ If significant changes have been made in either the water treatment facility and/or the water distribution system since the last PPC Survey, a request can be made for a new evaluation. ISO does not provide engineering advice on water supply improvements but they will provide guidelines on water supply demand for any new community development.

⬥ The insurance carriers for property protected by automatic sprinkler systems should be able to provide both supply and demand water curves for each protected property or HPR (Highly Protected Risk) property.

The following discusses how automatic sprinkler systems can be used to reduce needed fire flow demand.

Automatic Fire Sprinkler Systems

Cote and Linville state, "Automatic Sprinklers have been the most important single system for the automatic control of hostile fire in buildings for more than a century. Among the benefits of automatic sprinkler systems is the fact that they operate directly over the fire. Smoke, toxic gases, and reduced visibility do not affect their operation. In addition, *much less water is used* because only those sprinklers fused by the heat of the fire operate, especially if the building is compartmented." (4)

The above statement points out the fact that, of all the tools available to facilitate and promote fire protection, none offers such a wide variety of benefits to the building owner, developer, fire service, water authorities including conservationists, and the general public as does the widespread use of automatic sprinkler systems. One of the most important topics of this Chapter is to review a number of factors relating to the use of automatic sprinkler systems with special attention to their *effect* on required fire flows in building codes and the *effect* on ISO needed fire flows. Finally, some initiatives need to be considered in a further attempt to encourage the use of automatic sprinklers in both commercial and residential properties.

Advantages of Automatic Sprinkler Systems

While the insurance industry can be given much credit for promulgating and advancing the use of sprinkler systems in commercial properties, the cost and benefits of sprinkler protection in the United States industrial world were realized in the early 1900s. Warehouses and large industrial manufacturing facilities typically were protected by automatic sprinklers systems. This distinction often meant the difference between a company staying in business versus not being able to recover from a devastating loss.

Long before building codes required sprinkler systems, insurance companies gave substantial reductions in premiums for properties protected by automatic sprinkler systems.

Although automatic sprinklers were not recognized immediately for their life safety features, a trend was evident that indicted reductions in fire-related injuries and deaths when sprinklers were present. Currently, large area, highrise buildings, and high-value buildings and high density public buildings of all occupancy types signify the need for automatic sprinkler systems in all of the model building codes. As of 1990 the NFPA reported in the NFPA Journal that there was no record of a fire in 20 years that killed more than two persons in a completely sprinklered public assembly, educational, institutional, or residential complex where the sprinkler system was maintained and operated according to the original design. (5)

[Sidebar: The two deaths reported were from smoke inhalation in an area remote from the fire area.]

The Factory Mutual Research Corporation (FM) has documented some very interesting data on automatic sprinkler operations in industrial properties from 1987 though 1997. **Table 9-1** depicts some very important performance data in terms of number of heads operating, total number of fires based on heads operating, the cumulative number of fires, the percent of the total number of fires investigated–2,860, the cumulative percent and the estimated water flow for 28 minutes of operation, which was the longest period the sprinkler system is reported to have operated on any of the investigated fires. In other words, this total water flow was a maximum flow duration and, most likely, much less where 10 or fewer heads operated. These were ordinary hazards Group 3 properties where the sprinkler design was 0.20 gpm/sq. ft. and the area of sprinkler coverage was 100 square feet; this equates to a discharge of 20 gpm per head. (5)

Table 9-1: FM Study Sprinkler Performance Operations Study (1)

Number of Heads	Fires	Cumulative Number	Percent of Total	Cumulative Percent	Projected Flow Rate in Gallons Per Minute
1	797	797	28	28	20
2	520	1,317	18	46	40
3	291	1,608	10	56	60
4	221	1,829	8	64	80
5	147	1,976	5	69	100
6	146	2,122	5	74	120
7	74	2,196	3	77	140
8	77	2,273	3	80	160
9	44	2,317	1	81	180
10	55	2,372	2	83	200
11—15	172	2,544	6	89	220—300
16—20	100	2,644	3	92	320—400
21—25	49	2,693	2	94	420—500
26+	167	2,860	6	100	520—Not Estimated

Observations:

1) Note the outstanding fact that in 83 percent of the fires studied the fires were confined, controlled, and extinguished with 10 or fewer sprinkler heads operating. Assuming a design density of .2 gpm/ sq. ft. for ordinary hazard Group 3 property, the water flow would have been 200 gpm for a maximum of 30 minutes, or 600 gallons of water total. A fire of this size would have been confined to a fire area not *exceeding* 10,000 sq. ft. The water requirements from hose streams for a fire of this size in ordinary combustibles materials would have been in the range of 1,500 to 2,000 gallons of water.

2) From the above analysis it should be observed that automatic sprinklers also potentially reduce significant amounts of water damage. The cuts insurance losses two ways: 1) structural and content loss from the fire destruction, and 2) significant loss reduction from water damage.

3) From the FM study of the 2,860 fires in commercial and industrial complexes, it is important to note that even in the 6 fires where 25 or more sprinkler heads operated, the buildings of origin were saved from total destruction and could be rebuilt at a cost much less than rebuilding the entire structure. In all of the larger fires, the sprinkler heads confined the fire until the fire department could complete extinguishment with supplemental fire streams.

Water Supply Requirements for Sprinklered Properties

Under the ISO *Fire Suppression Rating Schedule*, needed fire flows represent a water flow demand based on calculations for structures and building *not protected* by automatic sprinkler systems. In recognition of the value and effectiveness of automatic sprinklers systems for the confinement, control, and extinguishment of developing hostile fires in structures and buildings, the needed fire flow requirement extends, in effect, 100-percent credit is given for all buildings within a *Graded Area* that are graded and credited as automatic sprinklered property.

The ISO-required water supply (i.e., not needed fire flow) consists of the flow required for operating sprinkler in the event of a fire and generally a hose stream allowance of 500 gpm for final extinguishment and overhaul of smoldering materials. The full range of the water supply for sprinkler heads will vary from 150 gpm to 1,600 gpm. These flows are dependent on the classification of hazard, whether the system is hydraulically designed or a pipe schedule systems in accordance with NFPA Standard No. 13 on *Sprinkler Systems*, the type of materials being stored, and the storage configuration. Under these limiting factors, the *maximum* amount of water needed for an insured building rated and codes for sprinkler protection is 1,600 gpm for the sprinkler heads + 500 gpm for the supplemental hose streams = 2,100 gpm; this is far different from needed fire flows that could range up to 12,000 gpm. (6)

Design Curves

NFPA 13 provides design criteria for automatic sprinkler system based on hazard level of the occupancy. These criteria are represented in design curves and in minimum water supply requirements. These design curves are used to determine the density required for various hazard classifications. (7) Density is defined as the flow required, in gpm per square foot, to be discharged over a selected area of operation. For example, if the density required is 0.10 gpm/foot squared and it is applied to an area of operation of 1,500 square feet, the minimum system demand, excluding hose streams, will be 150 gpm. These design curves indicate a range of basic system demands from 150 gpm for minimum light hazards to 1,600 gpm for extra hazard.

The required water supply for the sprinkler system developed for a specific fire risk from this curve may need to be increased to allow for two additional factors. To compensate for friction loss in the piping for pipe schedule systems, the normal design practice is to add 10 percent to the basic systems design. In hydraulically designed systems, discussed below, computer design models are used to select pipe sizes that will minimize friction loss for design flow and match the pipe design to available flow from water supply curves obtained for the building to be protected.

The second factor considers the hose stream allowance that needs to be included to permit the operation of inside and outside hose streams to complete extinguishment of fires in deep seated combustible materials. The hose stream allowance and fire flow duration are based on the hazard classification in accordance with the schedule in **Table 9-2**. (7)

It should be further noted that fire insurance carriers that write insurance for highly protected risks may have additional water demand requirements based on separate NFPA Standards or internal underwriting requirements. The high-rack storage of materials is a perfect example of this condition. When rack storage up to 20 feet in height is present, the design curves for ceiling sprinklers will increase to a maximum of 1,800 gpm for Class 5 unencapsulated commodities (NFPA 231C: *Covered Rack Storage of Materials*). If the rack storage is 25 feet high, the basic system will demand an increase to 3,250 gpm. (8)

Table 9-2: Hose Stream Allowance

Hazard Classification	Hose Stream Allowance Gallons Per Minute	Duration Minutes
Light	100	30
Ordinary—Group 1	250	60—90
Ordinary—Group 2	250	60—90
Ordinary—Group 3	500	60—120
Extra Hazard Group 1	500	90—120
Extra Hazard Group 2	1,000	120

Hazard Classification

It is apparent that, while sprinklered buildings are not included when determining the needed fire flow for insurance rating purposes, it is necessary to calculate the demand in order to ensure adequate water supplies to the sprinklered building. Average commercial occupancies, such as retail stores, offices, hotels, and institutional buildings, usually will fall into the category of ordinary hazard Group 1 occupancies. Warehouses and manufacturing occupancies will fall into ordinary hazard Groups 2 and 3. Occupancies in which there are highly flammable products or processes enlarge quantities will be classed as extra hazard.

The determination of special hazard classifications is defined in detail in specific NFPA Standards referenced at the end of this Chapter. These references need to be examined because automatic sprinkler system demands for potential sprinklered properties must be carefully calculated based on this data. Plans and calculations for such systems need to be submitted to the building department for the community, the fire marshal or fire protection engineer for approval and in many states the office of the state fire marshal for code compliance assurance, in addition to the insurance carriers engineering staff.

Residential Sprinklers Systems

There is an increasing requirement by local government, especially at the county level of government, to install residential type sprinklers in one and two family dwellings. One specific example is Prince Georges County located just north of the District of Columbia. There is now a sprinkler requirement for all new residential properties and rental space in existing homes. Since this ordinance was passed in 2001, there has not been a fatality in a home fire where residential sprinklers have been installed. Hostile fires in residential property with residential sprinkler protection have been confined to the room of origin or completely extinguished before the Prince George Fire Department arrived at the fire site where the system was functioning according to installation requirements. Two house fire receive moderate damage where the residential sprinkler system alarm device had been tampered with by the occupant or an electrician between 2002 and the end of 2005 when records were reported. In 2006, children were playing with matches in a bedroom and caught the bedding on fire. When the fire department arrived at the fire site, the fire was confined and extinguished by one operating residential sprinkler head. (9)

Records kept by the NFPA up to the beginning of 2006 has prompted the ISO recently to announce that it is reducing the needed fire flow for strictly single-family homes not over two stories in defined communities to a needed fire flow of 500 gpm regardless of exposure conditions. This is a *huge* reduction in required water supplies for closely built residential housing, and it will have a positive ripple effect on water supply requirements for both treated water and water supply pipe sizing.

Types of Sprinklers for Commercial Buildings

The required fire flow for automatic sprinkler systems is determined by applying the criteria in NFPA 13, *Standard for the Installation of Sprinkler Systems*. (6)

[It is extremely important to use the latest published edition of this Standard.]

Two types of sprinkler systems are in general use for commercial building:

1) Pipe schedule systems.
2) Hydraulically calculated or designed systems.

A pipe schedule systems is a sprinkler system that is designed using pipe size table with corresponding minimum water flow rates from NFPA 13. (6) A hydraulically designed system is a sprinkler system that is designed using existing or designated water supply pressure and calculating all water flow rates available for each size of piping in the system. Special sprinkler systems, such as deluge systems, occasionally are installed for special hazards. Today, it is almost essential that all *special types* of sprinkler systems be hydraulically calculated using computer modeling. (7)

Pipe Schedule Systems

The required fire flow for pipe schedule systems will vary from 250 to 1,500 gpm, based on the occupancy of the building. The required water supply at the needed residual pressure should be available at the base of the sprinkler rise to provide 15 psi at the highest elevation sprinkler head with sprinkler operating. In these types of systems the sprinkler pipe sizes are specified in the appropriate standard, and the piping is arranged like a "tree with branches" where all the branch lines and cross mains are dead ended; not an efficient water supply system, although these systems have been around for over 100 years. (7)

Hydraulically Calculated Systems

The required water demand for hydraulically calculated sprinkler systems will vary from 150 gpm up to several thousand gpm for business, entertainment, educational, religious, industrial, and governmental buildings, based on the area of the *largest* open area room (up to 5,000 square feet) and the occupancy. The required "fire flow for any given sprinkler systems" needs to be available at the base of the riser (alarm valve) at a sufficient pressure to deliver the required fire flow through the sprinkler piping to the *design area*, with a minimum pressure at every sprinkler head of 7 psi. NFPA 13 specifies a minimum size pipe and allows for looping the piping to cut down on the head loss in the piping system. The design of these systems is based on the calculated residual pressure available at the base of the automatic sprinkler system riser when delivering the required fire flow at the point of design. The calculated residual pressure is determined from the available fire flow data in the vicinity of the building to be protection and plotted on hydraulic graph paper.

[See Manual II on Water Supply Testing and Evaluation.]

The higher the calculated residual pressure for the needed sprinkler system flow demand, the smaller the sprinkler system piping that can be used. Such flow tests should be conducted when there is peak consumer consumption demand.

For economic reasons, these sprinkler systems are designed using the highest residual pressure at which the required fire flow is available. Unfortunately this information usually is needed before the building is built, and perhaps even before water mains are extended to the construction site. Therefore, it is important that the hydraulic data used accurately reflect the hydraulic conditions that can be expected in the vicinity of the proposed building. For both types of systems, the duration for which the sprinkler system must be designed to operate varies from 30 to 120 minutes, depending on the hazard of the occupancy and the ability to remotely control water flow.

It should be noted that a lowering of the normal pressure throughout a community or a general overall intentional lowering of the distribution system pressure can result in many formerly adequate sprinkler systems becoming inadequate. Hydraulically calculated sprinkler systems do provide an advantage in this regard for responding fire departments. It is now required that the design pressure and flow for the design area be recorded on a metal plate attached to the main control valve to the sprinkler system. The valve room should be accessible to the responding fire department in case of a fire. Observation of the pressure gauge during a fire can clearly indicate if the pressure on the system is being maintained. If the riser pressure drops below the demand pressure, the fire department can boost both the pressure and flow using fire department pumpers to supply hose streams to the fire department connection. The pressure at the sprinkler riser should be coordinated with the pump operator to maintain, but not exceed, the riser demand pressure. Too-high pressures will loft the water over he sprinklers making the sprinkler discharge less effective for controlling the fire.

Standpipes

Most local building codes require standpipes to be installed in any building five stories or 50 feet in height including buildings protected by automatic sprinkler systems. Standpipes are classified as follows: (7)

- Class 1: One 2-1/2-inch fire department outlet on each floor;
- Class 2: One 1-1/2-inch tenant hose station on each floor; and
- Class 3: Combined 2-1/2-inch and 1-1/2-inch outlets on each floor.

NFPA 14, *Standard for Installation of Standpipes and Hose Systems* covers the requirements for standpipes. (7) The water requirements for standpipes are 65 psi residual pressure at the highest outlet, with 500 gpm for the first standpipe and 250 gpm for each additional standpipe, to a maximum of 2,500 gpm. These requirements apply for Class 1 and Class 3 systems. A Class 2 standpipe system requires a total supply of 100 gpm at 65 psi residual pressure at the outlets. In fully sprinklered buildings, the water supply must satisfy the more stringent of two demands (stand-pipe or sprinkler system). Inside hose stream allowances for hose stations supplied from sprinkler systems (not necessarily Class 2 standpipe systems) of 50 gpm each, to a maximum of 100 gpm are added to the sprinkler water supply, but no minimum residual pressure is mandated. This inside hose requirement is included in the total hose allowance that was previously discussed.

Economic Advantages of Installing Automatic Sprinkler Systems

The *economic* advantages both to communities and individual property owners realized by protecting structural properties is *no small thing* as identified in the following examples and understanding related to property insurance and water supply.

Approximately 30 years ago, in the City of Fresno, California, the City Fire Marshal, the municipal planning department, the water department, and other municipal agencies joined in developing an innovative Mater Plan for Fire Protection. The plan was based around features of the insurance industry *Grading Schedule* combined with an integrated system of codes and ordinances. Prior to development of the master plan, the fire department budget represented approximately 13 percent of a total municipal budget of over $14 million dollars. At that time, the fire department was rated Class 1, but the water department was only rated a Class 3.

The codes and ordinances adopted as part of the new plan included a "dangerous-building ordinance" which gave the fire marshal authority to condemn property unsafe, a fire hazard, or unsanitary or that presented a clear and present danger to the community. Owners of condemned property had a choice of selling their buildings to an urban renewal agency or renovating the building to code standards, which included *mandatory automatic sprinkler systems*. In addition, all new construction required complete automatic sprinkler systems under the Federal Urban Renewal Agency Agreement with the city. (7)

At the time of this action, Fresno covered 21 square miles with a population of 115,000. The fire department maintained 68 career firefighters on duty at all times after full implementation of the mater plan development work for assistance in the sprinkler enforcement program. At the end of the program, 95 percent of all buildings in the urban renewal area were completed with automatic sprinkler protection. This area alone covered 40 square blocks of nonresidential property. Fire losses were reduced at least 22 percent a year after implementation of the sprinkler program, and the fire department retained a Insurance Class 1 rating.

All of this work and code enforcement results in more efficient distribution of municipal funds under the total municipal budget, and the water department classification rating changed from a 3 to a 1. The end result was reduced insurance premiums for all commercial classes of property in the city and added 10 percent to 30 percent insurance reduction for properties actually protected by automatic sprinklers

At about the same time period, a serious hardware store fire occurred in a community of about 14,000 population outside of Buffalo, NY. The building was brick wood-joist construction with no opening to exposures on both sides; one an insurance agency and the other a drug store. The fire was fought by the village volunteer fire department and three area fire departments operating under mutual aid. The building was essentially "gutted" but the exterior walls stayed intact on there was only smoke and water damage to adjacent occupancies.

The owner-occupant decided to rebuild and, with the advice of his insurance agent, installed a wet pipe automatic sprinkler system. The insurance agent showed the owner how the sprinkler system would pay for itself in 9 years through reduced property, content, and liability, and business interruption insurance. The owner also had the peace of mind that the likelihood of a second damaging fire was almost nil. This also had a positive impact on the water system by reducing the representative needed fire flow at that location to 500 gpm. (10)

In relation to the *Fire Suppression Rating Schedule*, every level of local government from the county, to the cities, to the towns, to the villages, need to understand that any commercial building that is coded and rated as sprinklered property for insurance purposes does not of and by itself have a needed fire flow requirement. The fire flow requirement is for the sprinkler system and the supplemental hose streams as indicated above in this chapter.

Based on the growing performance record of residential sprinkler systems, it is expected that more insurance reduction credit will be given for installed residential sprinkler systems in one-and two-family dwellings. In new developments that have a requirement for residential sprinklers, the needed fire flow is reduced to 500 gpm through the community. This has an impact on the need for new water mains and water treatment requirements.

The installation of automatic sprinkler systems in both residential and commercial properties is a win-win situation for both individual property owners and for municipal water supply systems.

Water as an Effective Fire Suppression Agent

Water has been used for fighting structural fires in America since the first fire department was organized in New Amsterdam, which was renamed New York City in 1664. Water is the agent of choice for structural fire suppression for many reasons: (11)

- The general ability of water from fire hydrants in communities that have a municipal water system.
- The general availability of water in suburban and rural areas without fire hydrant service from farm ponds, wells, small lakes, creeks, streams, and, in many locations larger rivers and lakes.
- These water sources can be used to provide *separate water supplies* for structural fire suppression, as discussed in Chapter 10. These water resources also can be used with dry hydrants to refill mobile water tankers and transport water to fire sites or, where feasible, by use of mobile pumpers that can be used to pump water directly to fire sites.
- Water is the least expensive of all the types of fire extinguishing agents currently available.
- It is available in a liquid state over a wide range of temperatures extending from 32 °F (0 °C), to 212 °F (100 °C), or the boiling point of water.
- Both treated and untreated water are generally considered safe to use in human and animal environments when *properly* applied through sprinkler heads or fire department nozzles.

🜩 It is most normally selected to control and extinguish structural fires because its nontoxic, noncorrosive, and is stable over a wide range of conditions. This means that when water H_2O is applied to a developing fire in ordinary combustible materials, the water does not break down into the basic elements of Hydrogen (H) and Oxygen (O), both of which would encourage fire growth.

🜩 Water can be applied to building fires while occupants are still in the building under approved fire-suppression techniques. This is not the case with some gaseous extinguishing agents, which may cause asphyxiation or adverse health side effects.

The Physical Properties of Water

It is essential that community officials, including the chief executive officer and staff, along with the water superintendent and fire chief understand the physical properties of water the permits the use of water as an effective fire extinguishing agent. The fundamental considerations include the following, documented by the NFPA. (2)

[Sidebar: The term "structure" is used in this Chapter to denote both unoccupied and occupied buildings for the sake of simplicity. Model building codes generally reference structures as being unoccupied and buildings as being generally occupied. This distinction will be made where the water application could be injurious to occupants still in a building when the initial fire attack is made on a developing fire.]

1) At ordinary room temperatures in a structure, water exists as a stable liquid. Water viscosity in the range of 24 to 210 °F remains constant, which allows water to be transported in water mains and pumped by fire department pumpers through fire hose.

2) Water has a high density, which allows it to be discharged from fire hose lines with nozzles and projected into structures on fire. Water's surface tension allows it to exist from small droplets to a solid stream.

3) The latent heat of fusion is the amount of energy required to change the state of water from a solid (ice) at 32 °F to a liquid. Water absorbs 143.4 Btus per pound in this process.

4) The specific heat of water is 1.0 Btus per pound. Therefore, raising the temperature of 1 pound of water 180 °F from 32 °F to 212 °F requires 180 Btus.

5) Water is an effective cooling agent because of its high *latent* heat of evaporation (i.e. changing water from a liquid to a vapor) which is 970.3 Btus per pound as will be described below.

6) Water expands in its conversion from a liquid state to a vapor state from 1,600 to 1,700 times the liquid volume. One gallon of water produces 223 cubic feet of steam. Therefore, it follows, that 1 gallon of water at room temperature applied to a fire and converted to steam (**complete conversion**) will have the following results.

   a) Heat required to raise the temperature of water to boiling: 212 °F - 68 °F (Average room temperature) = 114 °F. 114 °F × 1 Btu/lb × 8.33 (Weight of 1 gallon of water) = 1,200 Btu.

   b) Heat required to change water from a liquid to a vapor:
      970.3 Btu/lb × 8.33 = 8,083 Btu

   c) Total heat absorbed is 1,200 Btu + 8,083 = 9,283 Btu/gallon of water

The above concepts are extremely important to *effective* fire suppression as can be observed in the following example. A fire department hand-held hose stream discharging 100 gpm *could* have the potential capability to absorb 928,300 Btus per minute in a complete conversion from a solid state to complete steam conversions; this represents 100 percent effectiveness of using water to extinguish fire through both absorbing heat and oxygen displacement. This is realized by recognizing that the same fire department hose stream has the capability to create 22,300 cubic feet of steam in the conversion process.

To achieve this capability, water has to be applied in the form of a fog stream with finely divided particles in contrast to solid fire streams supplies by a tapered nozzle. A solid fire stream has been determined from numerous structural fire tests by the Western Actuarial Bureau to have only 10 percent or less of the cooling capacity of fog steams. In fact, 90 percent of the water applied from solid streams results in run-off with little or no extinguishing capability and may just do water damage inside a structure fire. (11)

All of this explains why the effective progress of a fire attack on a developing structure fire can be estimated and evaluated by observing the volume of smoke and condensing steam coming from the involved building. This has to be emphasized in the training of fire suppression personnel and especially in the training of tactical fire officers. Fire suppression personnel should be acquainted thoroughly with the characteristics of condensing steam and be able to evaluate the information it conveys on the fireground. Condensing steam can be observed and recognized instantly when projected against the sky. It is visible to the human eye in daylight or darkness. Observe a picture of the old steam locomotives when exhausting steam; the live steam is invisible as it comes from the exhaust valve. However, as it expands and loses some of its heat, the process of condensation begins and the expanded cloud of steam becomes visible. Condensing steam contains insufficient heat to cause physical injury. This has been used as a concern for not using fog nozzles to generate steam. To avoid any possible injury from steam generation at structure fires, the following three prerequisites are essential:

1) Fire suppression personnel needs to be trained to stay close to the ground or floor level where the attack is initiated since the steam will rise to the highest point of the enclosure.

2) Immediately after the initial fire attack, ventilation is required on the floor above the fire level or at the roof level for single-story structures.

3) Complete personal protective clothing is required for all fire service personnel making the fire attack; this includes self-contained breathing apparatus (SCBA) and the helmet ear flaps need to be down after the face piece is in place.

Above all, training fires need to be extinguished before attempting a real-world fire attack on a structure that could produce volumes of steam.

When all of this is in place, structural fire suppression effectiveness will significantly improve and the actual use of water to extinguish fires will be reduced for the communities that employ effective fire suppression which reduces water consumption records. As a final reminder, the following listing demonstrates the effective uses of water as a percent function for the cooling potential of 1 pound of water:

100 percent of efficiency—1,150 Btu/lb

75 percent of efficiency—863 Btu/lb

50 percent of efficiency—575 Btu/lb

25 percent of efficiency—288 Btu/lb

10 percent of efficiency—115 Btu/lb

Recall from above that a straight fire steam, whether from a hand hoseline or a master stream appliance has only about 10 percent of water efficiency. The above numbers clearly indicate the inefficiency of applying water by straight stream method, although this method may be necessary for coping with deep-seated fires and for overhauling fires. Water that is not used effectively in fire extinguishment often results in more water damage to the property than the fire itself.

The Latest Methods for Improving Structural Fire Suppression Effectiveness

Two fast-growing developments in structural fire suppression may have a significant impact on water demand needs for specific occupies and for initial fire attack in the near future.

First, is the use of water mist technology for special-hazard fire problems and for protecting one- and two-family homes.

Second, is both the introduction of Class A Foams and the technology to apply these foams for the reported dramatic improvement of confining and extinguishing developing structure fires to the room or area of origin.

Each of these approaches to improved structural fire suppression is examined separately.

Water Mist Fire Protection Systems

Since the early 1980s, there has been a very strong interest in the fire protection field considering the use of fine water sprays, better known now as *water mist*. The basic concept of *water mist* is as follows:

> Water is ejected from special nozzles at pressures ranging from 800 psi to 1500 psi to form sprays of finely divided water particles applied to generally confined areas where combustibles are present to extinguish a developing fire by the dilution of the air or oxygen content and by creating a water vapor through the space; maybe better defined as steam. The steam is generated from the rapid evaporation of the water droplets in the heated area surrounding the fire.

Instrumented tests of *water mist* application to room fires indicates that not only does it rapidly reduce oxygen content in the room which is needed for fire development but the *water mist* provides a very rapid cooling of the entire atmosphere which is important to confining the fire to the point of origin.

The concept of high-pressure fog for extinguishing ordinary combustible fires is not a new phenomenon. High-pressure fog nozzles were used on the "booster lines" of fire apparatus back in the 1950s. However, the lack of proper firefighting training resulted in many injuries to firefighters and therefore this type of equipment lost favor with the fire service. Installed *water mist* systems of today are an outgrowth of earlier technologies in structural fire suppression.

Fixed water mist systems is a subject of very high interest to all aspects of the fire protection field at the present time. In 2006, the Marriott Corporation was doing extensive studies of water mist applications to control and extinguish developing fires in residential occupancies to the item of origin or confine damage to the room of origin, first as a life safety measure, and second to minimize fire, smoke, and water damage to the room of origin. These sponsored tests were conducted at the University of Maryland as a cooperate effort between the Department of Fire Protection Engineering and the Maryland Fire and Rescue Institute. (12)

The term "*water mist*" implies a very fine water spray that remains suspended in air for an extended period of time. The term reflects one of the qualities of a spray (i.e., the drop sizes are very small) relative to raindrops, or to the spray from automatic sprinkler heads. For example, there are three basic characteristics that influence its effectiveness as an extinguishing agent:

1) The density of the spray, (mass of suspended water per unit volume of space).

2) The velocity with which it is delivered to the seat of the fire.

3) The quantity of water itself (which may contain dissolved additives to enhance suppression.

There are four characteristics of water mist for fires suppression purposes:

1) Drop size distribution.

2) Flux density.

3) Spray momentum.

4) Additives.

The current characteristics for Water Mist System are covered in NFPA 750, *Standard on Water Mist Fire Protection Systems*, 2003. (13)

Water Demand and Flow Duration

The water discharge rate for water mist systems is determined from full-scale fire tests for the type of hazard to be protected. Testing has been done on a wide range of fire scenarios, including shipboard cabin space for passenger ships through the Maritime Commission, machinery spaces, computer rooms, telecommunication centers and wire rooms, laboratory facilities of many kinds and both single- and multiple-family dwellings, along with hotels and motels. Research reports from these testing programs indicate flow ranges from 7 to 30 gpm for spaces occupied by humans. Machinery spaces may require flows up to 60 gpm. The recommended discharge period in NFPA 750 is 30 minutes. This implies that the needed water supply ranges from a minimum of 210 gallons to a maximum of 900 gallons for occupied spaces and a maximum of 1,800 gallons for machinery spaces. Compared to all other types of fire extinguishing methods with water, this is a tremendous reduction is water supply for fire control, confinement, and extinguishment. Remember, the fire tests conducted include water for final extinguishment.

Special Considerations for Residential Fire Protection

The following material addresses applications of *water mist* that applies specifically to the residential property classification. It should be understood that significant research on *water mist* for residential properties is ongoing. One important example is the cooperative effort of former fire chief and now Director of Corporate Fire Protection, Sonny Scariff, Marriott International, Inc.; Deputy Chief Michael Love, Montgomery County, Maryland, the Maryland State Fire Marshal's Office; Gary Briese, Executive Director for the IAFC; Dr. James Milke, Department of Fire Protection Engineering, University of Maryland; which formed a partnership to do *water mist* testing on room fires in hotels, motels, multiple occupancy residential properties, and single-family dwelling. (14) Other research is on-going at the NIST, Underwriters Laboratories, Inc. (UL) FM, and private organization including "Hi-Fog" Water Mist Fire Protection, Inc.

In the United States, the greatest loss of life from fire continues to be in the single-family home. Many fire protection leaders feel that this Nation has reached the limit to which we can **dramatically** further reduce the numbers of those injured and killed by fire each year in our country. (14)

Using current approaches to fire protection, it does not appear that this Nation will decrease the number of lives lost in residential fires much further than reported by NFPA in 2005. What may be even worse is that the numbers may actually increase, given that our fire services are being called upon to accomplish so much more in the post 9-11 environment. The American fire service is now an emergency services delivery system that is the first responder to any emergency eventuality. In order to make a major shift in the status quo, it is essential that new concepts and applications be applied to protect residential occupants from hostile fires.

Of further importance, in 2005, 82 percent of the fire deaths in America occurred in the home, to say nothing about the associated property loss from these deadly fires. Based on data from both the NFPA and the USFA: (14)

- On the average, single-family residential fires kill 9 people every day in America.
- On the average, 45 residential fires are reported every hour across the Nation.
- Single-family residential-type fires account for nearly 6 billion dollars in direct property damage each year since 2000 (with the dramatic increase in hosing costs this number will climb).
- Fires in homes are a constant threat from the largest cities to the smallest towns, in communities of all economic levels, in both newly constructed dwellings and in dwellings built many, many years ago.

In response to this carefully documented fire problem in residential properties, many communities are passing ordinances that require the installation of residential sprinkler systems in **new** residential construction. No documentation has been found on any requirement for retrofitting residential sprinklers in homes. Retrofit water mist fire protection systems just may be the answer for the proper protection of residential properties without affecting the municipal water system demand. The Marriott Corporation, along with the Hi Fog Corporation, are currently (2006) conducting live burn tests in a single-family dwelling at the Maryland Fire Rescue Institute facility, College Park, MD, to study different retrofit *water mist* technologies relating to nozzle type and placement nozzle pressures, and the amount of water to be stored in high-pressure cylinders, types of alarming devices in relation to a wide range of fire scenarios in the living room, kitchen, dinning area, and bedrooms while providing fire protection for all occupied space. This work should form the basis for writing a standard on the design, installation, and testing of *Water Mist Systems for Residential Occupancies*. (14)

Class A Foam Fire Protection

Ever since World War II, a number of different additives to water have come on the market in hopes of improving fire protection in general. Of significant importance has been the development of several foam concentrates to be used with water for the extinguishment of flammable liquid or Class B fires. More recently, Class A foam agents have been developed for ordinary combustible materials, with the objective of improving structural fire suppression effectiveness using both hand-held fire streams and used in conjunction with automatic sprinkler systems. The fire service literature claims that Class A foam makes fire streams up to 100 percent more effective than plain water. A larger view of the fire protection literature, including some full-scale fire tests and laboratory tests at the NIST in Gaithersburg, MD, point out that there are a number of variables associated with determining the effective use of Class A foam with fire streams. The percent of increased effectiveness in the use of Class A foam with automatic sprinkler systems is definitely positive but, again, is dependent on the type and arrangement of the combustible fuel loading.

At the beginning of 2006, there appears to be a significant amount of confusion about Class A foam concentrates and their application to structural fire suppression.

A first consideration is that Class A foams do not contain chemical fire-retardant compounds. Class A foam has it's origin with forest and wildland firefighting since the early 1980s. Today, this Class of fire-retardant chemicals is considered by many forest officials to be the agent of choice when making wildland and forest air drops in advance of an oncoming flame front(s).

Key Concept: The forestry service Class A fire retardants **are not** the same as the synthetic Class A foam agents for structural fire suppression. These two Class A foam agents should not be confused.

The NFPA definition of a chemical fire retardant is "any substance, with the exception of water, that by chemical action reduces the combustibility of fuels or slows down their rate of combustion." A Class A foam concentrate approved for structural fire suppression is classified as a fire suppressant." As a fire suppressant, UL and other approval agencies state that Class A foam concentrate is mixed with water and directly applied to *burning fuels* to provide both blanketing and wetting actions. Furthermore, Class A foams for structural fire suppression do not *chemically inhibit fuel ignition.* (15)

Listed below are some basic characteristics of Class A foam agents for structural fire suppression:

1) There is promotional material on Class A foam that appears to confuse firefighters concerning the finished foam as a chemical fire retardant. The issue revolves around the facts concerning the possible delay in ignition by the efficient wetting of fuels that absorb moisture, such as a dwelling with unsealed wood siding that is threatened by fire exposures. Indications are that one should not expect the Class A foam approved for structural fire suppression to act as a fire-retardant agent. Fire retardants generally are applied to fuel surfaces before the flames approach. This they continue to retard fuel ignition even after the water mixed with the chemicals evaporates from fuel surfaces. The retardant action continues to occur for a specific time period based on the specific agent after application. (15)

2) Once Class A foam for structural fire suppression has been applied and has evaporated, ignition can reoccur if final extinguishment has not been achieved and proper overhaul procedures are not carried out. There is little evidence to indicate that Class A foam solutions approved for structural fire suppression will chemically inhibit fire ignition.

3) Another issue concerns single applications of Class A foam solution to ordinary combustibles. Tests indicate that after a single Class A foam application, the water evaporates, **or** the heat content of the burning material boils off internal moisture, and erupts in flames after the materials reach ignition temperatures. This is why it is essential to either reapply the foam solution to the burning material or after "flame knockdown" complete the extinguishing process with water.

Despite these considerations, there are a lot of real-world success stories about using Class A foam for structural firefighting, and almost all new fire engines are being equipped with foam proportioning systems for Class A foams or compressed air foam systems. Journal articles and books are promoting the use of Class A foam for structural fire suppression. The following points appear to present the best current understanding on the use of Class A foam as predicted in the fire service literature:

● Class A foam delivered from 1-1/2- to 2-inch hoselines with fog nozzles have demonstrated very effective knockdown of developing fires and single and multiple family homes, barn fires, and small commercial building fires.

● Depending on the area of the fire involvement, the initial fire attack has extinguished the fire. However, re-applications are needed to prevent reignition, which indicates that below the foam surface ignition temperatures still exist. Foam followed by a water application appears to be the most effective way of controlling the fire and extinguishing the fire.

🔹 Less water is used when Class A foam is employed for the initial fire attack, except where the building is fully involved upon arrival of the first fire apparatus. Some of the best success stores come from volunteer fire departments that protect rural areas without a municipal water supply.

🔹 A comprehensive training program involving fires at a training center or buildings that are to be town down is a **must** for firefighters who are going to use Class A foam.

🔹 The ISO has recognized the use of Class A foam by giving credit for carrying the type of foam on fire department pumpers.

References

1. Insurance Services Office, Inc. "Effective Fire Protection–National Concern." Insurance Services Office, Inc., 545 Washington Blvd., Jersey City, NJ 07310-1686.

2. _____. *Fire Suppression Rating Schedule*. Insurance Services Office, Inc. (ISO), 545 Washington, Blvd., Jersey City, NJ 07310-1686.

3. AWWA-17. *Installation, Field Testing, and Maintenance of Fire Hydrants*. American Water Works Association, 6666 West Quincy Ave., Denver, CO 80235.

4. Cote, A.E., and J.L. Linville, eds. *Fire Protection Handbook*, 16th ed. Quincy: National Fire Protection Association, 1986.

5. Isman, Kenneth E. *Automatic Sprinkler Systems*. Ch. in NFPA *Fire Protection Handbook*, 18th ed. Quincy: National Fire Protection Assocation. 2000.

6. NFPA 13: *Standard for the Installation of Sprinkler Systems*, 2002 Edition.

7. American Water Works Association—M-31. *Distribution System Requirements for Fire Protection*. American Water Works Association, 6666 West Quincy Ave., Denver, CO 80235, 1989.

8. NFPA 232: *Standard for the Protection of Records*. Quincy: Author.

9. Unpublished material by the Marriott Corporation, Washington, DC, 2005.

10. National Board of Fire Underwriters, S.I.B. *Value of Automatic Sprinkler Systems, 1968*.

11. Layman, Loyd. *Attacking and Extinguishing Interior Fires*. Boston: National Fire Protection Association, 1952.

12. Marriott Corporation, Unpublished Literature, 2005.

13. NFPA 750, *Standard on Water Mist Fire Protection Systems*. Quincy: Author, 2003.

14. Mawhinney, Jack R. *Water Mist Fire Suppression Systems*, NFPA *Fire Protection Handbook*, 19th ed. Vol. I. Quincy: NFPA, 2003.

15. David and Colletti, *The Rural Fire Fighting Handbook*. Lyhon's Publishing.

CHAPTER 10 SEPARATE WATER SYSTEMS AND EMERGENCY WATER SUPPLIES

Part I

During the middle 1800s and up to 1915 the United States experienced what often has been referred to as the "Age of Conflagrations." Major fires in large cites and some smaller cites during this period literally destroyed the commercial areas of the city and, in some cases, the whole community. The more significant fires occurred during the San Francisco earthquake, the conflagration-type fire that ripped through Chicago, the downtown business fire right after the turn of the century in Baltimore, MD, and the shipyard fire that spread throughout much of Charleston, SC. The list is rather extensive as depicted in **Table 10-1**.

Table 10-1: Famous Conflagrations in the United States Occurring Prior to 1915

Date	Location	Property Destroyed	Dollar Loss*
1835	New York, NY	Buildings covering 13 acres	$ 15,000,000
1845	Pittsburgh, PA	1,000 buildings	$ 3,500.000
1849	St. Louis, MO	425 buildings, 45 steamboats	$ 3,500,000
1851	San Francisco, CA	2,500 buildings	$ 3,500,000
1861	Charleston, SC	Ships, docks, and business district	$ 10,000,000
1866	Portland ME	1,500 buildings	$ 10,000,000
1871	Chicago, IL	17,430 buildings–250 people killed	$ 168,000,000
1871	Pestigo, WI	17 towns destroyed by forest fires–1,052 people killed	$ No loss data
1872	Boston, MA	776 building–13 people killed	$ 75,000,000
1874	Chicago, IL	Waterfront buildings	$ 5,000,000
1889	Seattle, WA	Commercial and residential properties	$ 5,000,000
1889	Spokane, WA	Commercial and residential properties	$ 6,000,000
1189	Boston, MA	52 commercial buildings–4 killed	$ 3,600,000
1889	Lynn, MA	Commercial and residential properties	$ 5,000,000
1892	Milwaukee, WI	Commercial district	$ 6,000,000
1900	Hoboken, NJ	Piers and steamships–326 killed	$ 4,600,000
1901	Jacksonville, FL	Commercial and residential properties	$ 11,000,000
1902	Paterson, NJ	525 commercial buildings	$ 5,500,000

1904	Baltimore, MD	80 city blocks in the central business area	$ 50,000,000
1906	San Francisco, CA	28,000 buildings–earthquake/conflagration	
		452 killed or missing	$ 350,000,000
1908	Chelsea, MA	3,500 buildings, 11 killed	$ 12,000,000
1911	Bangor, ME	267 buildings, 2 killed	$ 3,189,000

* Fire losses are estimated for the year of the conflagration type fire. These numbers do not reflect growing economic values. (1)

As with most large-scale fires in cities and communities, the water system in each incident was overtaxed. Water pressures dropped with the number of hosestreams placed into operation to the point where fire streams were not effective and it is reported that hosestreams were shut down to establish a perimeter around the fire to keep it from spreading. The fires in Baltimore and Chicago destroyed the entire commercial districts, while the earthquake that started fires in San Francisco all but destroyed the whole city because the earthquake disrupted the entire water system.

In response to the water-supply-crisis during these fires, the idea of establishing independent water supplies that would supply special fire hydrants to support higher pressures so that hose-line could be fed off from the hydrants without pumping equipment to provide required pressures on nozzles at the end of hoselines. These individual water systems, which are separate from the domestic treated water system, are still in existence in some major cities. A brief description of two current systems is provided below.

1. Baltimore, Maryland.

This city essentially rebuilt its individual fire water system some 30 years ago by placing the pumping station at the lowest level to the block-square central fire station located a half mile from the waterfront. The water supply comes from two 20-inch suction pipes leading from different locations in Chesapeake Bay, and travels underground to the pumping equipment in the fire station. Trash racks are used to filter out any debris from the bay water. Water leaving the pumping station is treated with injected chlorine as a disinfectant, but the water does not meet EPA standards for drinking water. A 20-inch loop main extends around the high-value district of downtown Baltimore with interconnecting secondary feeders along major streets. Special high-pressure hydrants are primarily located at the intersection of many streets and painted a distinguishing *orange* color in contrast to the red fire hydrants on the domestic water system. In the older part of the downtown section of the city, fire hydrants are located at least every 1,000 feet on the secondary feeders. The pumping station has the capability to deliver up to 10,000 gallons of water a minute at pressures up to 300 psi. (2)

The system pressure is kept at 60 psi to monitor any possible leaks or breaks in the water system. On routine first-alarm fires, water is supplied from fire hydrants on the domestic water system to supply mobile pumpers. The high-pressure system is activated on demand by the highest ranking fire officer at a fire site or emergency. The system was most recently use when a CSX freight train traveling through Baltimore derailed in a tunnel with a hazardous cargo and caught fire. (3)

2. New York, New York.

The old high-pressure individual water system in New York City was replaced in the 1960s with a different type of individual water supply approach to augment the domestic water system used for normal fire-suppression operations. The water supply for New York City is unique in that the water source comes from reservoirs and lakes in the Catskill Mountains, 80 to 120 miles from the city limits. Water is transported by gravity pressure through 10-foot aqueducts to treatment plants north of the city. Large water mains are installed in a grid system through the streets of New York. However, the piping system is very old, so the pressures are kept relatively low, and building pumping systems are used to develop the higher pressures required for the multistory buildings. Due to the very old building construction going back to the mid-1600s in what is referred to as "loft buildings," the FDNY has to fight some very challenging fires and, in doing so, the loss of life on this fire department is quite high.

A 40-inch main now runs parallel to Broadway in New York City, and is cross-connected every 10 blocks to the Hudson River and the East River. On the river fronts at these cross-connector are located a battery of fire hose connections. Either mobile fire pumpers or fire boats can pump water from the two rivers as required to pressurize the individual water supply which, in turn, supplies special "steamer" or 6-inch outlets on fire hydrants mounted on this special pipe network. (4) At the time of the 9-11 disaster at the twin towers in New York, the domestic system supplied an estimated 8,000 gpm for firefighting while the individual water system provided another 10,000 to 12,000 gpm for an approximate total of 20,000 gallons of water being pumped at the fire site every minute during and after the collapse of the towers. This gives another dimension to the terrific nature of this event. (5)

A Current Need for Individual Water Systems for Fire Protection

The water supply below the aquifer is diminishing at a rapid rate in population growth areas. On a community water system in southern Maryland, the required well depth was 57 feet in 1960. Forty-six years later the well depth is 759 feet, and the water system is running out of water. The cost of treating water to meet EPA water-quality standards is rising exponentially. It appears that a significant portion of these costs, along with meeting higher consumer demand plus building-code-required fire flows, and the ISO needed fire flows is becoming an extremely costly consideration for taxpayers. The key consideration here is that water for public fire protection over 500 gpm should be studied and costed out as a *separate* water supply for considerations as an intelligent alternative to the way water is being supplied to communities today. Water supplies in excess of 500 gpm to handle the initial attack on developing structural fires do not have to be treated water that meets EPA water quality standards. It only needs to be screened to prevent trash, leaves, and other debris from entering the intake to supply pumps; a literature search to review Federal and State water supply requirements for fire protection indicates that only chlorination is required to prevent microorganisms from forming the pipe lines. This reduces the needed fire flow on the municipal water system, it adds reliability to the water supply for fire protection, it conserves water for domestic use, and, economically, it should reduce taxes and property/casualty insurance rates. The design and installation of individual water systems could be a Federal works project reducing the financial burden of these systems on local governments.

Basic Design Concepts for Individual Water Supplies for Fire Protection

Three separate types of water systems have been identified on the East Coast. Two of these systems are operational, and the third is in the planning and design stage. The communities involved have requested that only generic information be presented at the request of the municipal Attorney's Office. A conceptual scheme for each system is provided.

1) **System 1:** The following system is located in a northeast State. The community population is about 7,500 people and is located on a substantial river that typically has a very good year-around flow rate in million gallons per day past the suction point for the individual water system. This suction point is upstream of the water treatment plant. The community has two old mill buildings, a fair-size business district and a Federal enclave of buildings, plus residential areas. The community also serve a farming region and has a rather large commuter population to a larger city.

   The community dates to the Civil War era and the buildings are either brick wood-joist or frame construction; the old mill buildings, even though modernized, still are heavy beam and timber construction with a wood roof and wood siding requiring a needed fire flow of 6,500 gpm. Fire flow tests in the area indicate an available fire flow of 3,550 gpm under an average daily consumption rate. Therefore, there is a significant deficiency in meeting the community need fire flow at several locations.

   In order to increase the available fire flow without improvement to the water treatment facility and the distribution system, the fire department sought the advise of a civil engineer for the County Department of Soil and Water Conservation. Plans were made to use the nearby river as a water supply to feed a loop main around the business district with separate fire hydrants to be painted blue in contrast to the regular red fire hydrants. Fire hydrants were located 500 feet apart around the loop. A pump house was located 200 feet from the volunteer fire station, and a concrete pad was installed so that two mobile fire department pumpers also could draft water from the river and pump water into the loop main as a reliability factor to the pump house, or to pump directly into the loop main using 6-inch connections with the necessary reducers for the supply hose being used. Check valves were installed to prevent backfeed to either the industrial pump or to the mobile fire department pumpers. Weatherproof bourdon gauges where installed so the residual pressures could be monitored at the supply connection. The objective was to provide the best possible volume at any fire hydrant on the loop at 20 psi residual pressure.

   The stationary pump selected was a diesel-driven industrial water pump that could deliver 2,000 gallons of water a minute at 50 psi header pressure. Separate flow tests after the pump house installation indicted a flow of 2,000 gpm at 20 psi by the old mill buildings, and 1,850 gpm at 20 psi by the middle school on the back side of the pipe loop. These flow were increased at the same pressures by almost 1,000 gpm when a 1,500 gpm mobile pump also was pumping into the system. Therefore, in a severe emergency, the community water system could potentially provide 3,500 gpm for fire suppression and the individual water supply could provide an additional 3,000 gpm for a total of 6,500 gpm within the business district of the community. Using large-diameter hose and relay operations, the two individual water systems should be able to meet all needed fire flows throughout the community including the Federal enclave. The community has not been regraded by ISO since the individual water supply was placed in service.

2) **System 2: An Individual Primary Feeder Pipe to Supply Fire Hydrants in Florida**

In central Florida there is an island that is located between the Atlantic Ocean and the Intracoastal Waterway. For a stretch of about 12 miles in a highly developed resort area of highrise buildings, there is an installed independent 10-inch water main that runs South from a pumping station that takes water from the Intracoastal Waterway. Nonfrostproof fire hydrants with only a pumper connection are located every 1,000 feet on this pipeline. The pumping station has two 1,500 gpm pumps that have discharge pressures of 100 psi. One pump is electric drive and one pump is diesel drive. The discharge piping is arranged and gated to provide for independent or parallel pumping into the water main. Water flow tests at the time of the installation indicate that a minimum of 3,000 gallons can be delivered at 20 psi residual pressure on any three consecutive fire hydrants located on the independent water source. This supply is in addition to the normal municipal water supply availability. (7)

3) **System 3: An Independent Water Supply with Two Supply Points to Protect Two Adjacent Communities**

On one of the eastern Great Lakes there are two communities that have a common political jurisdiction line. The larger of the two communities have a population of about 27,000 and is located adjacent to the lake; this will be identified as Community #1. There are several small industries in this community that are dependent on the lake water. The municipal water supply comes from the lake, and fire protection is provided by a combination career and volunteer fire department.

Community #2 is about 3 miles inland from the lake and is home to a moderately sized university and supports an agricultural surrounding area with retail stores, shops, banks, restaurants, and hotel/motels. This community supports a volunteer fire department, and the public works department has charge of the water supply system. Water supply comes from a manmade reservoir with a high-level dam located 2 miles from the village limits.

Both communities have underground water supply piping that is at least 100 years old on many residential streets and the needed fire flows in each community is less than 50 percent adequate in the high value districts and in the perimeter residential areas. The second community has a special water supply problem. During the harvest and canning season, water flows in the main business district have been recorded as low as 135 gpm at 20 psi residual pressure by a fire protection engineer.

The two communities have embarked on a joint water supply program that is both unique and based on a cooperative agreement between the two communities to prove significantly improved water supplies for fire protection and improve the reliability of treated water for each community in case of drought conditions for Community 2 or ice blockage on the intake lines to the treatment plant for Community 1. The basic planning has been completed by a professional engineering firm with a fire protection engineer on the staff. A cost development study is in progress, and funding is being sought through grants to implement the plan.

[Sidebar: The two communities combined experienced from a low of 75 structural fires per year to a high of 111 fires per year from 1995 to 2005.]

The cost development study will assess the potential impact on property insurance rates, needed improvement costs to the existing water supply treatment facilities, water distribution system piping costs, and cost revenues from supplying nontreated water for municipal and industrial uses, and possibly for supplying water to be treated for consumer consumption between the two communities during peak-demand periods and for emergency purification in case of a water system disruption.

The general plan calls for frostproof hydrants with pumper connections to be located every 500 feet in the business, commercial, and industrial areas of each community and 1,000 feet in the residential areas where the existing fire flow from the treated water system is less than 500 gpm at 20 psi during periods of maximum daily consumption. The details for this smaller piping layout had not been completed at the beginning of 2006, and probably will not be available until the funding source(s) are in place for the primary piping system. For this manual, it is the basic concepts that are important as a means of meeting fire flow requirements while reducing the needed expansion of two water treatment plants and enhancing the water distribution piping system. The following points identify the significant features of this plan:

- In Community 1, a separate water pumping station would be supplied by lake water where the intake is 10 feet below the 50-year lake drought index as measured by a hydrologist. Two 2,000 gpm pumps would be located in the pumping station that can be controlled by the water treatment plant operator. One pump would have an electric driver and the second pump would be diesel driven with remote start capability. In order to monitor the pipe condition, 25 psi static pressure would be maintained on the piping system at all times. Note: This includes all of underground piping for both communities up to the two control valves.

- Community 2 would provide untreated water from the reservoir through two primary feeders to the identified control valves. The Public Works Department that provides staffing 24/7/365 would activate these valves as requested by the water treatment facility in Community 1 based on actual flow demand. Water from the reservoir would enter the pipe network at about 80 psi, and the pressure profiles would balance through the individual distribution system.

While this is a conceptual plan at this writing, the potential for transferring the demand for fire protection water supplies higher than 500 gpm and for reducing the need for treated water supplies for a variety of uses, plus the potential for emergency water supplies to be discussed later as a function of Homeland Security deserves a second look. Even if the cost-benefit analysis is not promising for the experimental program, it may be more feasible for larger population groups. This is an example of thinking "outside the box" from current standards and textbooks to solve both current and future water demand issues.

Part II: Emergency Water Supplies

The concept of emergency water supplies as referenced by the AWWA pertains to implementing procedures to provide potable water in the event that a municipal water supply system is in a failure mode for whatever reason.(8) A failure mode could apply to a section of a community because of a pipe failure, or to the entire community if there is a drought and the water supply source dries up. These are two example of water system unreliability where emergency water supplies need to be transported or otherwise provided for a community population, or the population has to be evacuated to a community that has an ample water supply.

After 9-11 the conceptual framework of emergency water supplies has taken on a different perspective. A water system could be knocked out by a terrorist attack that physically disrupts the water treatment facility, the water distribution system, or biologically contaminates a community's water supply source, which is the equivalent of having a "dried up" water supply. Likewise natural disasters such as floods, earthquake, and severe weather can create disruptions to water supplies.

In response to such events, it is essential that municipal government officials, water supply superintendents and their staffs, the fire administrator/fire chief and senior fire officials along with consulting engineers plan for and initiate an emergency water supply delivery program to transport potable water to specific areas of a community or to the entire community.

Three independent methods of transporting potable water for human consumption are identified below. There are other techniques that are being evaluated at the present time, especially where the community is located in the vicinity of a salt-water source. Water superintendents will learn more about this from future AWWA publications.

1) **Method 1:** Provide an underground secured vault interconnection between two different potable water systems. Examples include adjacent communities, or communities with a separate potable water system provided by a college or university, a State or Federal government facility including a military base, or a private water company that services the public school system in some counties; obviously there are other examples. The AHJ may require backflow prevention devices so that one system cannot backflow to the other system after the emergency condition has been corrected.

2) **Method 2:** A separately valved fire hydrant is placed at the remote point on Water System A near an adjacent water system. The adjacent Water System B also has a separately valve-controlled fire hydrant within 50 feet of Hydrant A. A 1,000- to 1,500-gallon mobile fire department pumper is arranged to pump from one system to the other; this can be from Hydrant A to Hydrant B or Hydrant B to Hydrant A. Discharge pressures should not exceed 40 psi unless approved by the water supply superintendent for the receiving water system, or a staff member. Overpressurization could cause another pipe failure.

3) **Method 3:** Some county fire and emergency services organizations have a converted a used "milk tanker" pulled by a semi-tractor as an emergency vehicle that can fill and transport 8,000 gallons of water to individual buildings (e.g., schools, nursing homes, hospitals, apartment buildings, etc.) in case they lose their municipal water supply or a well system goes dry. Because of the operating costs involved this type of water transport is best implemented where there is a county-level office of emergency services and a fire station or an emergency service training facility that can house the mobile tanker for what should be considered infrequent use. Maintaining the sanitation of tankers that are seldom used can be problematic. One alternate is to contract with firms who transport dairy products or other potable liquids to supply these emergency support services.

There is still another approach to providing potable water to buildings or areas where either the municipal water supply is out of service, or wells run dry. The Navy especially, and now all branches of the military services including the National Guard have portable water-purifying equipment that uses reverse osmosis to purify nonpotable water from streams, lakes, rivers, and even salt water from the oceans. An environmental company on the West Coast has built a prototype fire department pump that has a combined reverse osmosis and water-treatment package so that the pumper can take draft from essentially any water source and pump either nonpotable water for fire suppression or potable water to be delivered through a special polyvinyl hose to fill water cans for domestic use.

References:

1. NFPA *Fire Protection Handbook.* 12th ed. Quincy: National Fire Protection Association, p. 1-56.

2. American Insurance Association (A.I.A.) S.I.B., 1969.

3. "Baltimore Sun" newspaper article., Sept. 16, 2001.

4. American Insurance Association (A.I.A.) S.I.B., 1969.

5. "Special 9-11 Report." *Fire Engineering,* Jan. 2002.

6. American Water Works Association-M-31. *Distribution System Requirements for Fire Protection.* American Water Works Association, 6666 West Quincy Ave., Denver, CO 80235, 1989

7. Florida State Fire Marshal's Report, Mar. 1996.

8. American Water Works Association-M-31. *Distribution System Requirements for Fire Protection.* American Water Works Association, 6666 West Quincy Ave., Denver CO 80235, 1989.

www.ingramcontent.com/pod-product-compliance
Lightning Source LLC
Chambersburg PA
CBHW081126170526
45165CB00008B/2573